3 COMMANDO BRIGADE
HELMAND
ASSAULT

BOOKS by EWEN SOUTHBY-TAILYOUR

Military Histories and Biographies

Falkland Islands Shores

Reasons in Writing: A Commando's View of the Falklands War

Amphibious Assault Falklands: The Battle for San Carlos

Blondie: A Life of Lieutenant-Colonel HG Hasler DSO, OBE

The Next Moon: A Special Operations Executive Agent in France

HMS Fearless: The Mighty Lion

3 Commando Brigade, Helmand

Nothing Impossible: A Portrait of The Royal Marines 1664–2010
[Editor]

Fiction

Skeletons for Sadness

Reference

Jane's Amphibious and Special Forces

Jane's Special Forces Equipment Recognition Guide

3 COMMANDO BRIGADE
HELMAND ASSAULT

EWEN SOUTHBY-TAILYOUR

EBURY
PRESS

1 3 5 7 9 10 8 6 4 2

First published in 2010 by Ebury Press, an imprint of Ebury Publishing
A Random House Group company
This edition published 2011

The information in this book is believed to be correct as at 1 September 2010,
but is not to be relied on in law and is subject to change. The author and
publishers disclaim, as far as the law allows, any liability arising directly or
indirectly from the use, or misuse, of any information contained in this book

The Random House Group Limited Reg. No. 954009

Addresses for companies within the Random House Group can be found at
www.randomhouse.co.uk

A CIP catalogue record for this book is available from the British Library

The Random House Group Limited supports The Forest Stewardship
Council (FSC), the leading international forest certification organisation.
All our titles that are printed on Greenpeace approved FSC certified paper
carry the FSC logo. Our paper procurement policy can be found at
www.rbooks.co.uk/environment

Mixed Sources
Product group from well-managed
forests and other controlled sources
www.fsc.org Cert no. TT-COC-2139
© 1996 Forest Stewardship Council
FSC

Designed and typeset by seagulls.net

Printed in the UK by CPI Cox & Wyman, Reading, RG1 8EX

ISBN 9780091937768

To buy books by your favourite authors and register for offers visit
www.rbooks.co.uk

CONTENTS

Dedicated to the memory of those members of

3 Commando Brigade, Royal Marines

who were killed during *Operation Herrick 9*

Corporal Philip Smith
42 Commando, Royal Marines

Marine Jamie Hutton
42 Commando, Royal Marines

Trooper James Munday
Household Cavalry Regiment

Rifleman Yubraj Rai
2nd Battalion, The Royal Gurkha Rifles

Lance Corporal Neil David Dunstan
United Kingdom Landing Force Command Support Group,
Royal Marines

Marine Robert Joseph McKibben
United Kingdom Landing Force Command Support Group,
Royal Marines

Patrolman Enyat Ullah
Afghan Territorial Force

Colour Sergeant Krishnabahadur Dura
2nd Battalion, The Royal Gurkha Rifles

Patrolman Aman Ullah
Afghan Territorial Force

Marine Alexander Lucas
45 Commando, Royal Marines

Marine Tony Evans
42 Commando, Royal Marines

Marine Georgie Sparks
42 Commando, Royal Marines

Private 1st Class Dan Gyde
Danish Battle Group

Private Jacob Gronnegard Gade
Danish Battle Group

Marine Damian Jonathan Davies
Commando Logistic Regiment, Royal Marines

Sergeant John Henry Manuel
45 Commando, Royal Marines

Corporal Marc Birch
45 Commando, Royal Marines

Lance Corporal Steven Jamie Fellows
45 Commando, Royal Marines

Lieutenant Aaron Leslie Lewis
29 Commando Regiment, Royal Artillery

Rifleman Stuart Winston Nash
1st Battalion, The Rifles

Sergeant Jacob Moe Jensen
Danish Battle Group

Private Sebastian la Cour Holm
Danish Battle Group

Private Benjamin Davi Sala Rasmussen
Danish Battle Group

Corporal Robert Christopher Deering
Commando Logistic Regiment, Royal Marines

Lance Corporal Benjamin Whatley
42 Commando, Royal Marines

Corporal Liam Elms
45 Commando, Royal Marines

Serjeant Chris Reed
6th Battalion, The Rifles

Marine Travis Mackin
United Kingdom Landing Force Command Support Group,
Royal Marines

Captain Tom Herbert John Sawyer
29 Commando Regiment, Royal Artillery

Corporal Danny Winter
45 Commando, Royal Marines

Acting Corporal Richard Robinson
1st Battalion, The Rifles

Corporal Danny Nield
1st Battalion, The Rifles

Marine Darren James Smith
45 Commando, Royal Marines

Lance Corporal Stephen Michael Kingscott
1st Battalion, The Rifles

Marine Michael Laski
45 Commando, Royal Marines

Corporal Tom Gaden
1st Battalion, The Rifles

Lance Corporal Paul Upton
1st Battalion, The Rifles

Rifleman Jamie Gunn
1st Battalion, The Rifles

Lance Corporal Christopher Harkett
2nd Battalion, Royal Welsh

Corporal Dean Thomas John
The Royal Electrical and Mechanical Engineers
1st The Queen's Dragoon Guards

Corporal Graeme Stiff
The Royal Electrical and Mechanical Engineers
1st The Queen's Dragoon Guards

Staff Sergeant Henrik Christian Christiansen
Danish Battle Group

Marine Jason Mackie
Armoured Support Group, Royal Marines

Lance Corporal Robert Martin Richards
Armoured Support Group, Royal Marines

Raised in urgent, clouded days, the Commandos hardened themselves for battle by sea, land or air, in which nothing was certain except the hazards they would have to face. To them, danger was a spur, and the unknown but a challenge.

HM Queen Elizabeth, the Queen Mother,
unveiling the Spean Bridge Commando Memorial, Achnacarry
27 September 1952

LIST OF MAPS

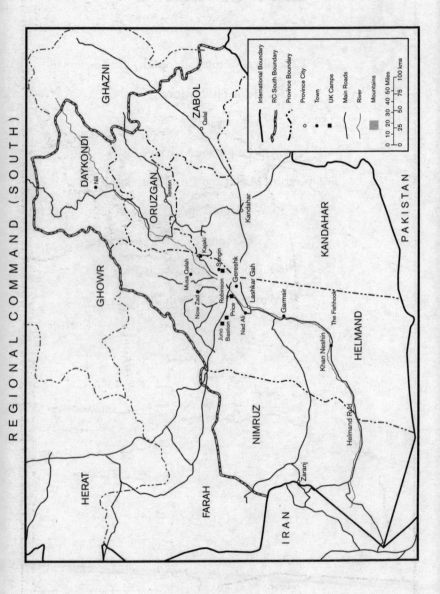

REGIONAL COMMAND (SOUTH)

International Boundary	
RC South Boundary	
Province Boundary	
Province City	○
Town	●
UK Camps	■
Main Roads	
River	
Mountains	

0 10 20 30 40 50 Miles
0 25 50 75 100 kms

GHAZNI

ZABOL

Qalal

DAYKONDI

Nili

ORUZGAN

Tereen

GHOWR

Kandahar

KANDAHAR

PAKISTAN

Kajaki

Musa Qalah

Sangin

Gereshk

Now Zad

Robinson

Price

Lashkar Gah

Garmsir

Juno

Bastion

Nad Ali

The Fishhook

Khan Neshin

HELMAND

HERAT

NIMRUZ

Helmand R♭d

FARAH

Zaranj

IRAN

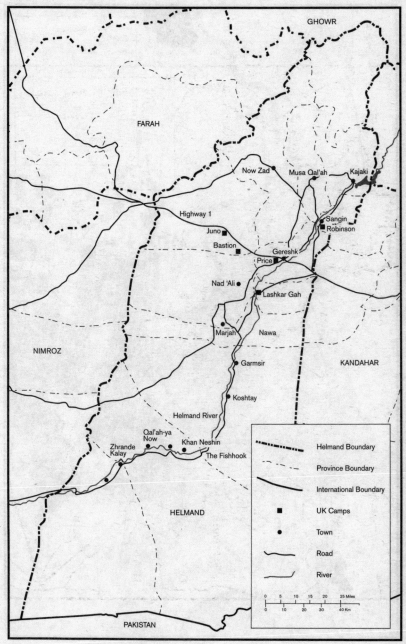

LASHKAR GAR

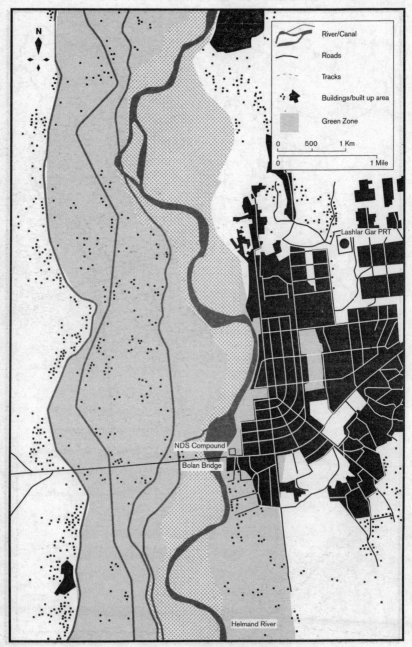

OPERATION SOND CHARA

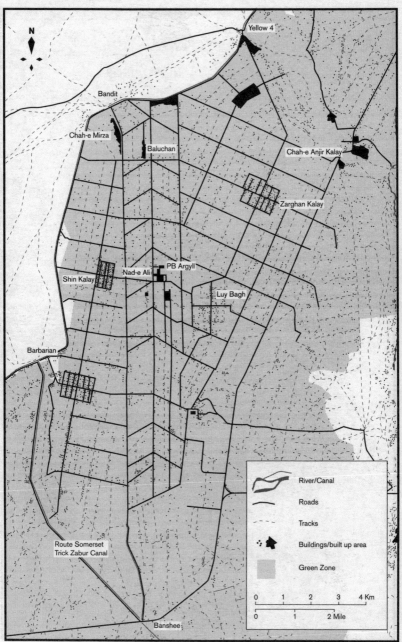

Yellow 4

Bandit

Chah-e Mirza

Baluchan

Chah-e Anjir Kalay

Zarghan Kalay

Shin Kalay

Nad-e Ali

PB Argyll

Luy Bagh

Barbarian

Route Somerset
Trick Zabur Canal

Banshee

	River/Canal
	Roads
	Tracks
	Buildings/built up area
	Green Zone

0 1 2 3 4 Km

0 1 2 Mile

OPERATION MARLIN
AND PATROL BASE JAKAR

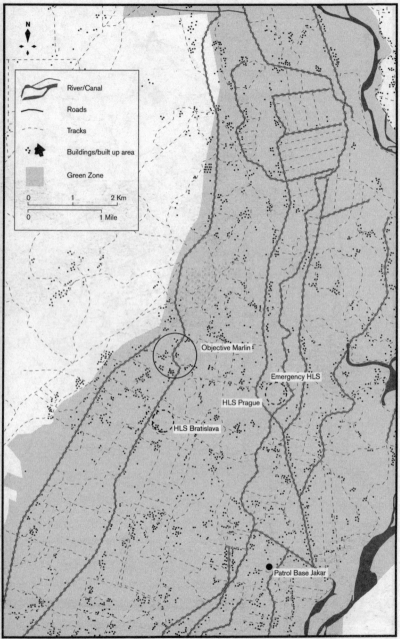

SANGIN VALLEY

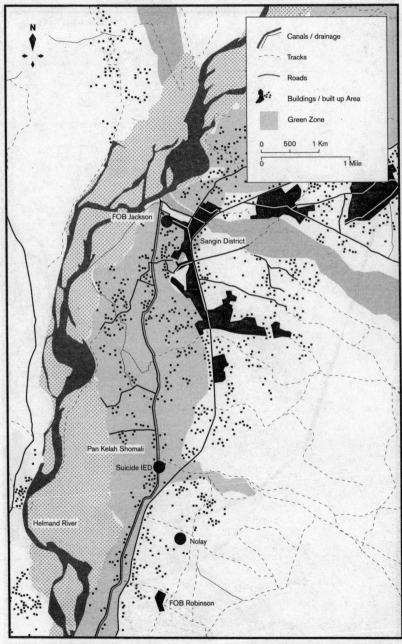

FOB Jackson

Sangin District

Pan Kelah Shomali

Suicide IED

Helmand River

Nolay

FOB Robinson

Canals / drainage

Tracks

Roads

Buildings / built up Area

Green Zone

0   500   1 Km

0   1 Mile

N

OPERATION DIESEL

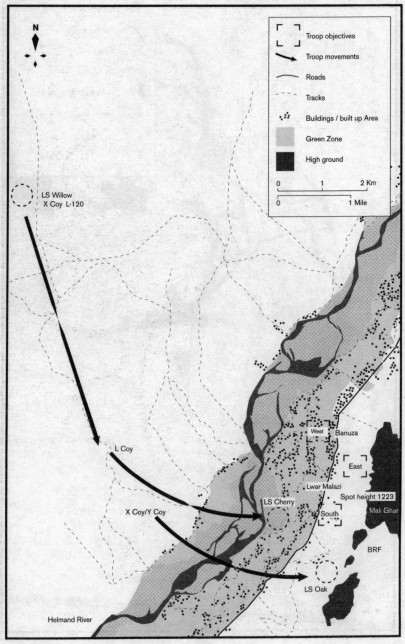

LS Willow
X Coy L-120

L Coy

X Coy/Y Coy

LS Cherry

West Banuza

East

Lwar Malazi

Spot height 1223

Mali Ghar

South

BRF

LS Oak

Helmand River

N

0   1   2 Km

0   1 Mile

Troop objectives

Troop movements

Roads

Tracks

Buildings / built up Area

Green Zone

High ground

THE FISHHOOK / AABI TOORAH 2 B

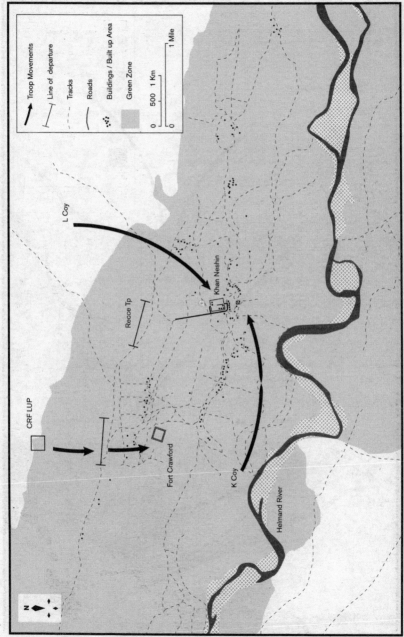

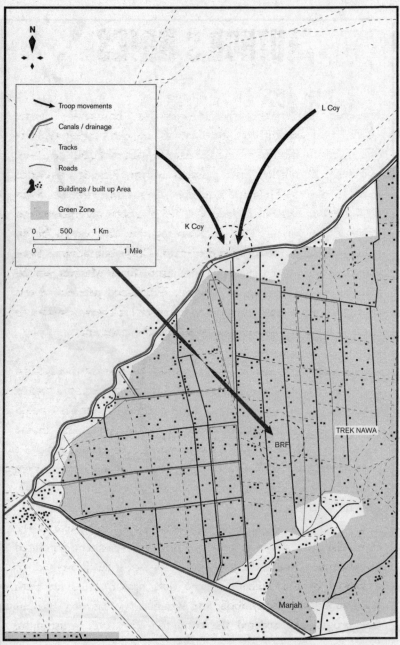

AUTHOR'S NOTES

This is the story of the Royal Marines' 3rd Commando Brigade, which was at war for seven long, exhausting, exhilarating, dangerous, successful months during the Afghan winter of 2008 to 2009.

Having chronicled brief passages of the Brigade's previous sojourn across Helmand during *Operation Herrick 5* in *3 Commando Brigade, Helmand*, it had not been my intention to repeat the experience but ... Knowing that this would be an account aimed not so much at the serving Corps but rather at those who might consider serving, Headquarters, Royal Marines, agreed with Ebury Publishing's suggestion. Unfettered permission was then granted for me to interview whomever I felt was necessary to obtain a good cross-section of tales from the front line.

The Brigade's mission in 2008 was not to destroy the Taliban across all six provinces of Regional Command (South), but to defeat the insurgents in order that good Afghan governance could flourish under the authority of the Provincial Governors, for the undoubted benefit of the local communities. If guaranteeing that governance had to involve engaging with Taliban fighters then that had to be, but it was not a prime aim during the Commando Brigade's watch.

Local governance in Helmand, assisted and supported by the Provincial Reconstruction Team and British Government-appointed stabilisation advisers, built schools, for so long prevented or destroyed by the Taliban; opened new health clinics for both sexes; monitored the symbolic and significant wheat seed distribution programme; secured the conditions for voter registration;

constructed the first new, custom-built police station in the country; and inaugurated a province-wide radio station.

Men from the Brigade destabilised the Taliban across the south, unearthed significant hoards of arms, ammunition and IED-making equipment, while always uncovering and then destroying vast quantities of drugs and the wherewithal for processing the raw opium, harvested from poppies.

Yes, Taliban were killed in their hundreds but unlike some – more often the 'embedded media' who seldom see the wider picture beyond the limited horizon of a battlefield – the well-trained serviceman knows that the success of a military operation tends to be in inverse proportion to a large body count on either side. The removal of the insurgency's leaders by whatever means is important but, in the long term, the reconciliation of the Taliban's locally bred and poorly trained 'foot soldiers' was more so.

As with all British brigades operating in southern Afghanistan, this is a tale of courage, endurance and supreme professionalism, all backed up by sheer dogged determination and patience – needed for success at both the social and political levels – as well as by the point of the bayonet.

In the space available it has been possible to follow only a very small selection of the actions seen through the eyes of the junior commanders and marines who took part. The tale must be regarded as merely a series of selected vignettes illustrating the marine commando at his best; doing what he does best, and doing what he likes doing best (or, second best!). Thus this story is very definitely not history; nor, equally definitely, is it complete.

I have made little mention of the daily grind experienced by, for instance, the marines and soldiers of 45 Commando Battle Group, including those of the 2nd Battalion Royal Gurkha Rifles. These men patrolled daily out of their besieged patrol and forward operating bases to face down the Taliban intent on preventing legal governance, progress and reconstruction. Often a thankless task, it was undertaken skilfully and courageously, as was the task assigned to 1 Rifles as the operational mentoring and liaison team battalion. Their work remains fundamental to the future of Afghanistan and yet it receives little or no recognition. One day, I hope.

The Commando Logistic Regiment with its near-superhuman Combat Logistic Patrols under permanent IED threat also more than deserve their own story to be told. Again, one day.

Nor has there been room to mention, let alone describe, the dozens upon dozens of stand-alone operations that marked the whole of the Commando Brigade's tour. All these were supported by the gallant – more than gallant – sappers of 24 Commando Regiment, Royal Engineers; the stalwart gunners of 29 Commando Regiment, Royal Artillery; and the valiant technicians of the 3rd Battalion Royal Electrical and Mechanical Engineers, who simply never stopped even when under fire and extreme pressure. The Scimitar light tanks of the Queen's Dragoon Guards were in almost continual vigilance and attendance throughout: always there, always ready, always on time and always on target.

It is a story that should also have included, had space allowed, even more of the hundreds of heroes in Helmand other than those whose exploits are almost daily in the news: the 'unsung' gunners, sappers, signallers, medical and logistical teams who all deserve full public praise, as do the fixed- and rotary-wing air crew and their ground staff; the tanks and vehicle crews; the post-room workers, clerks, electricians, plumbers, cooks, cleaners and laundry staff and indeed everyone who is needed to keep the equivalent of a small town and the people who operate out of it – and in a war zone at that – active, healthy, clean and safe.

*

The marines of the Commando Brigade brought with them to Helmand that extra edge that only thirty-two weeks of the toughest and longest infantry training course in the world can offer: the stamina that manifested itself time and time again as they completed lengthy 'yomps' across the deserts, carrying close to their own body weight, then through the green zone and, most dangerous of all, between miles of hostile compounds.

This remarkable endurance was best demonstrated by 42 Commando's L Company who, during *Operation Sond Chara*, yomped for over sixty kilometres, not only fighting three major battles on the way but also under continual threat of ambush and IED. If that was not sapping enough, much of the operation was conducted in what were accurately described as Somme-like conditions. That they tragically lost just one man – Lance Corporal Ben Whatley, a 'tough, uncompromising commando' who, characteristically, was leading his men into action when he was killed – is added testament to their personal battle skills, extreme fitness, exemplary field craft, training and dedication. When the whole of *Operation Herrick* comes to be written, L Company's 'sixty-kilometre yomp in contact' will be held up as one of the supreme performances of all who will have fought in Helmand Province.

But there were others, of course: 45 Commando Battle Group's remarkable achievements during the eight-phase *Operation Diesel* were the ultimate examples of painstakingly planned and meticulously executed commando helicopter assaults at night. Supported by a lengthy Brigade Reconnaissance Force decoy patrol across the desert and through the mountains, the whole was a classic example of a commando forces, all-arms raid.

The conduct of the hastily-brought-together Battle Group Centre South during *Operation Sond Chara*, with the Princess of Wales's Royal Regiment, and Estonian Infantry in support, will be studied for years to come as a prime illustration of intelligent, counter-insurgent operations executed in concert with local authorities.

Nor did any ground operation take place without the painstaking efforts of the Information Exploitation Group, whose tireless, never-ending input to the intelligence picture was fundamental to every successful operation – that there were no failed operations is a worthy testament to the IX Group's proficiency: this was cerebral soldiering at its very highest level.

A similar accomplishment was that of 42 Commando on *Operations Aabi Toorah 2B* and *2C*, supported by the Commando's own Reconnaissance Force and, as with 45's operations, by the British Army's Apache attack helicopters; RAF Chinooks; Royal Navy Sea Kings, Lynxes and Harriers; and the United States' Sea Stallions, Black Hawks, Cobras and B-1B bombers.

And no mention of Helmand would be complete without praising the Ammunition Technical Officers – the ATOs, or bomb-disposal experts – of the Royal Logistics Corps, without whose selfless exposure to danger from IEDs many, many more British and civilian lives would have been lost.

Photographs are always a tricky subject. Every professional image captured by an official military photographer is deemed to be Crown Copyright. Having received over 5,000 photographs – many from private sources – it has not been possible for me to credit them all other than to record their 'official' status, although where it has been possible to identify the individual photographer I have done so. Those official photographers whose names are known are: Petty Officer (Photographer) Dave Husbands of the Command Support (IX) Group; Leading Airman (Photo-grapher) 'Gaz' Faulkner of 42 Commando and Leading Airman (Photographer) Nick Tryon of 45 Commando. Others whose photographs will be found between these covers include Colonel Rory Copinger-Symes of Brigade Headquarters; Major Andy Watkins, attached to 1 Rifles; Major Chris Haw of the Brigade Reconnaissance Force; Major Marcus Taylor of the Commando Logistic Regiment; Lieutenant Commander Rob Stevens of the Commando Helicopter Force: Captain Orlando Rogers of 42 Commando; Corporal 'Bugsy' Malone of 45 Commando and Marine Jeff Saayman of 42 Commando.

The maps have, once again, been drawn by Tim Mitchell – a Royal Marines signaller for ten years – of Tim Mitchell Designs. I am as delighted with the results here as I was with those he drew for *3 Commando Brigade, Helmand*.

Inevitably the list of 'thank-yous' is a long one. With the now-mandatory permission of the MOD needed prior to any interview being held during 'service time' (and having to pay for that time), almost everyone listed below was happy to forfeit an off-duty, stand-easy lunch, evening or weekend in order to assist in compiling this brief snapshot of a handful of operations during *Herrick 9*. Some helped by merely saying 'yes' to my requests for information before pointing me in the right direction and then letting me get on with it. I am extremely grateful to everyone whose names – in no specific order – are recorded here, and to those whose help had to go unacknowledged in the text. I have not been able to incorporate direct, attributable quotes but where relevant I have used extracts from interviews to enhance the story (ranks are those believed to have been held at the time of interview):

Major General Andy Salmon whose initial encouragement was paramount to the project. Brigadier Gordon Messenger, Colonel Martin Smith, Colonel Rory Copinger-Symes, Lieutenant Colonel Al Litster, Major Chris Haw, Captain Chris Gardiner and

WO1 (RSM) Marc Wicks all kindly gave me the initial interviews on which so much could then be built.

Captain John Hillier's help is always invaluable coupled with permission to copy, quote and plagiarise from his marvellous journal, *The Globe and Laurel*: without this assistance so much colour would be missing. Lieutenant Commander Paula Rowe (Chief Media Ops during Herrick 9 and now of the MOD) offered patient understanding and advice that has been more than appreciated. The Brigade Media Team led by Sergeant Warren Keay-Smith pointed me in various directions, including the vital map store and the ever-helpful Sergeant Martin Cochrane.

Lieutenant Colonel Andrew McInerney and Captain Nick George of the Command Support Group Royal Marines (now renamed as 30 Commando Information Exploitation Group, Royal Marines) were tireless in their support.

Lieutenant Colonel Jim Morris managed to fit me in over lunch before accompanying the then Secretary of State for Defence to Headley Court, while Lieutenant Colonel Oliver Lee, his PMC and the officers of 45 Commando made me welcome at Arbroath before a series of interviews cleverly orchestrated by Lieutenant Simon Williamson, Royal Navy. Major Nicky Jepson's huge help with *3 Commando Brigade, Helmand* has now been over-matched by even more help for this book. Those others in '45' who also gave their time so generously include: Major Rich Maltby, Major Rich Parvin, Major Niki Cavill, Captain Neill McCurry, Sergeant Dave Taylor, Corporal John Ballance, Corporal Wayne Harrison, Corporal Ian Bishop, Warrant Officer 2 Kev Cheeseman, Warrant Officer 1 (RSM) Steven Shepherd and Colour Sergeant Matthew Wildgoose.

Lieutenant Colonel Charlie Stickland, Major Oli Coryton and the RSM Warrant Officer 1, Dom Collins, set the ball rolling at

42 Commando, followed by Major Adam Crawford, Warrant Officer 2 Adrian Webb, Marine Stuart Pinkawa, Corporal John Prideaux, Sergeant Adrian Foster, Lance Corporal Chris Bedford, Marine Jeff Saayman, Marine Ben Tomlinson, Marine John Fulton, Corporal Jonathan Mosley, Sergeant Dermot Buckley, Marine Pete O'Hanlon, Major Reggie Turner, Major Rich Cantrill, Major Jules Wilson, Sergeant Noel Connolly, Sergeant Richie Guy and Warrant Officer 2 Ed Stout.

Lieutenant Colonel Haydn White – late of the Helmand Provincial Reconstruction Team – provided, over a long Saturday, excellent source material during a fascinating interview.

Major Andy Watkins offered a closely detailed account of events on 12 March 2009 when Naval Medical Attendant Kate Nesbitt won her Military Cross: it was a privilege to meet her, too.

Ali Nightingale of Ebury Publishing has shown great patience, as has my eternally long-suffering agent, Robin Wade, whose own fascinating military career might, one day, deserve scrutiny! Belinda Jones has helped edit the manuscript in a most masterful manner for which I am exceedingly grateful. Thanks to the very tight deadline, my family and especially my wife, Patricia, have once again borne my prolonged absences with fortitude – and, possibly, with gratitude!

Opinions expressed that are not directly attributable are mine, for which I take full responsibility. The views of those interviewed do not represent those of the Ministry of Defence.

Finally, I pay tribute to all those wounded in the course of *Operation Herrick 9* and especially to those forty-four men who paid the ultimate sacrifice, of whom twenty were Royal Marines. We shall remember them.

PROLOGUE
'CONTACT – WAIT OUT.'

A swift diagnosis was needed as Kate leapt into the small trench. From her first glance she knew instantly that it was going to be touch and go. The round appeared to have hit Jon's right shoulder webbing, ricocheting through the top of his left lip, through his teeth and tongue and exiting through his neck, blowing away much of his left jaw and the back of his tongue as it did so.

The danger lay with Jon's inability to breathe. As she knelt by his head, a new burst of fire smacked into the ground around their position.

'Get down, Kate. On your bloody belt buckle.' Captain Foster tugged at her webbing as, kneeling, Kate struggled to rid herself of her medical Bergen. She couldn't do that lying down.

Jon List was barely alive: drifting in and out of consciousness, spluttering on his own blood and with a failing breath rate ... Kate tried desperately to find a pulse.

Operation Herrick 9 was executed in Afghanistan between September 2008 and April 2009 by the marines of the 3rd Commando Brigade, Royal Marines; 1st Battalion The Rifles; and various supporting arms. For many who took part, it was not their first time in the maelstrom that is Afghanistan; for most, it was not their first time in battle. This is their story.

The Royal Marines' mission during *Operation Herrick 9* was to secure and stabilise earmarked areas of responsibility, and then work in conjunction with the area's Provincial Reconstruction Team to extend and deepen the influence and affect of the Afghan government across Helmand Province.

Provincial Reconstruction Teams (PRTs) are made up of military officers, diplomats and reconstruction experts, whose efforts set in place the infrastructure and facilities for good, democratic governance in unstable areas. This was achieved by continuing the training and mentoring of the Afghan army and police force – training that is carried out by international forces' operational mentoring and liaison teams (OMLTs) – which in turn enabled a variety of other projects to be in place during each brigade's tour of duty.

The Commando Brigade's aim was not to destroy the Taliban, but to break its hold on local territories. By becoming a force for good in areas of responsibility, as opposed to the presence of menace and threat, hearts and minds would be convinced, and the rebuilding of a democratic Afghanistan could continue and strengthen.

One young woman who had not been in Afghanistan before but who, on the morning of 12 March 2009, was into her sixth month of *Operation Herrick 9*, was Able Seaman Class 1 Kate Nesbitt. A blonde, twenty-two-year-old Royal Navy medical attendant over 680 kilometres from the sea, Kate was one of two medics working at Patrol Base Jaker, Helmand Province.

As bullets screamed overhead during their latest contact, Captain John Foster, the second in command of C Company, 1 Rifles, was desperate to make himself heard above the eardrum-piercing crack and no less dangerous buzz of the Taliban's incoming 7.62 bullets.

'For Christ's sake, Kate, stay down! Stay down! You're fucking being shot at!' He tugged at the naval medic's Osprey body armour, but she shook him off.

'I know!' Kate yelled too, trying to raise her voice above the whizz of the now-familiar Afghan 'bees'. 'I know I'm being fucking shot at but I'm trying to save Listy's life here! Now bloody well hold on to this …!'

The patois of combat does not change over the years.

Kate Nesbitt should not have been there at all. Having worked in care homes as a teenager, her desire to remain in medicine was satisfied when, in December 2005, she was accepted into HMS *Raleigh* for basic naval training before being posted to Aldershot for fourteen months' medical training and a final four months in HMS

Nottingham to wind everything up. As her father and brother were Royal Marines, she then volunteered for service at the Commando Training Centre, Lympstone. Here she received the signal posting her to 3 Commando Brigade for *Operation Herrick 9*.

When the number of medics from all branches of the armed forces required for Afghanistan had increased, Kate had been pleased: with the Royal Marine Commando Brigade about to take the lead in action in Helmand, that number would be predominantly from the naval service.

Life at the Commando Training Centre prepared Kate well for her forthcoming deployment: she provided medical cover for commando courses in all their phases, often spending up to two weeks in the field with the young recruits, especially on Dartmoor and at Sennybridge. And as part of her preparation for her forthcoming operational tour, Kate joined the Commando Logistic Regiment's Medical Squadron at Chivenor in north Devon.

Kate arrived in Helmand on 26 September 2008 and the next day moved into Camp Shorabak. She was to have been sent to Sangin and the headquarters of Battle Group North with 45 Commando but as this was her first operational tour, it was decided that a quieter introduction to life on the front line would be better and so she was appointed to 1 Rifles, the operational mentoring and liaison team battalion.

For the first months, Kate was based in Camp Shorabak, the base for the Afghan National Army's 4th *Kandak* (battalion) of the 3rd Brigade of the 205th Corps, about 30 kilometres north-east across the desert from Lashkar Gah. As one of six medics mentoring their opposite number in the ANA, Kate found herself in the frustrating position of being confined to the base: as a woman, no one was quite sure how the ANA 'warriors' – the name for their private soldiers – would accept being treated medically by a woman, so, until R&R or until opinions altered, Kate was restricted to Camp Shorabak. She managed to escape on a few occasions, however, but due to the operating procedures in place at the time, she was always required to have a British male medic with her as a chaperon.

More permanent than the supposedly temporary but comparatively comfortable Camp Bastion, which was about twenty minutes' drive away across the sand, rubble and scrub of the *dasht*, Shorabak was linked to Camp Tombstone – itself an adjunct to Camp Bastion. Kate's sickbay was a Nissen hut with a wall dividing it in two; the other half being the US military's Ops room.

With its concrete buildings and air conditioning, Shorabak was certainly not as spartan as Patrol Base Jaker but it was definitely an Afghan base, with only rudimentary facilities for the British. These included a basic cricket pitch that, although it had seen better days, was still in use. Britain vs Afghanistan were closely fought games; when on the losing side, the Afghans would complain that they could not afford proper cricket balls with which to practise.

Separate kitchens served good English food prepared by cooks from 1 Rifles, who even managed theme nights each Friday, to which Americans from their nearby Camp Tombstone would be invited. Kate was surprised to find that there was a laundry manned by members of the base staff, computers for internet access, telephones, a fully equipped gym and, luxury of all luxuries, flushing lavatories. Kate and her colleagues slept in cot beds – under which they kept their weapons – with duvets.

There were no Christian compounds of worship but the camp boasted the newest mosque in the province. It was, though, in a war zone.

*

During one of Kate's chaperoned patrols, she first met Andy Watkins, a Royal Marines major on secondment and commanding C Company, 1 Rifles, who mentioned that his medic was due on leave and that he was going to be short of medical cover. Jumping at the implied offer, she was accepted on a two-week trial.

Luckily for Kate, those two weeks extended to four weeks. During that time, she covered a number of minor operations when her physical fitness, although severely tested, was never found wanting. But fitness to operate under battle conditions was just as important as muscle and stamina: to think and act quickly, clearly and calmly while under fire or when exhausted, and often both at once. No quarter was given to gender on the battlefield as Kate had earlier discovered during endless realistic training exercises at the Commando Training Base at Lympstone. There she had to carry a stretcher as swiftly and as far as the next man, and prove herself better at administering first aid than the men. When the day ended, she had to dig her own shellscrape and erect her own 'bivvy', all the while offering a medic's sympathy and care.

Luck continued to play its part, as when C Company's medic returned from his mid-tour R&R, the only other medic of C Company went home for good, and the post was permanently hers.

Little did she know it, but Kate's current patient, Lance Corporal Jon List, should not have been out in the thick of things, either. On being posted to Afghanistan, he was due to join the battalion's intelligence cell but had asked to see his company commander, declaring that he wanted to fight.

'Don't we all?' had come the reply. 'You'll be in a fight, all right!'

Six months before deploying to southern Afghanistan as a now-permanent manoeuvre unit within the Commando Brigade,

1 Rifles had been earmarked to oversee and train the Afghan National Army (ANA) by carrying out operational mentoring and liaison team duties.

In earlier years, the work of the inadequately equipped OMLTs in Afghanistan had been conducted with difficulty due to the daily, intense fighting that was taking place. This was to have consequences: during *Operation Herrick 5*, carried out from November 2006 to April 2007, for example, 45 Commando had taken over from a unit who had been unable to concentrate on this aspect due to the demands of day-to-day fighting. As a result, they had not enjoyed a comfortable ride with the ANA. Thus, as the importance of the operational mentoring and liaison teams' duties became apparent over the months that followed, the role was afforded additional attention, with units more carefully chosen and diligently trained. One of *Operation Herrick 9*'s main aims was to concentrate on mentoring and liaison team issues.

The task of mentoring and training the Afghan National Army, not only in Helmand Province but across the whole of Afghanistan, is not an enviable one. It is a difficult, dangerous but vital job, recognised as such because it is widely agreed that once the Afghan military and police forces are at full recruitment level, and trained, the International Security Assistance Forces (ISAF) will be able – as an experienced American general put it – to pack up and go home. The day is some way off but is coming, the main sticking point now being the Afghan National Police, who are proving less easy to train due to tribal allegiances and the temptation of easy money made from drugs and low-level extortion.

So in any 'normal' society, it is not the Afghan National Army warriors that the civilian population want to see on their streets, any more than Britons would want to see infantry patrolling Britain's city streets. In a well-run society, it is the role of the civil authorities, and not the army, to maintain law and order: the Afghan National Army may well reach its target to the satisfaction of NATO, but it is the police force's recruitment target figure that is, in the long term, the more important, and will be the more difficult to achieve.

It is not a task for the unprepared or the unprofessional but, under the redoubtable command of their commanding officer, Lieutenant Colonel Joe Cavanagh, 1 Rifles was certainly equal to their mentoring task. In April 2009, they were eventually able to hand over to their successors an Afghan *kandak* that was not only trained for battle but one that was also, crucially, battle hardened.

By 12 March 2009 – the day that was to become so important to Kate Nesbitt – the harsh conditions of Patrol Base Jaker, deep in the Taliban heartland of Nawa, were all too familiar to both the 4th *Kandak*, and to 1 Rifle's C Company.

Under the command of Major Andy Watkins, C Company had deployed to Helmand Province on 22 September 2008 and was soon established in the Garmsir district centre in Forward Operating Base Delhi. It was in Garmsir's ruined agricultural college that a captain in the Royal Irish Regiment, Doug Beattie, had been relieved from his near-siege circumstances and heroic command in September 2006 by the Information Exploitation Group of the 3rd Commando Brigade: since Beattie's and his soldiers' action, Forward Operating Base Delhi and the notorious Jugroom Fort, a few kilometres south along the Helmand River, had remained in ISAF hands, growing in size and becoming more secure as time passed. Growing, too, had been the number of outlying patrol bases that had been established to help widen the footprint of Battle Group South.

Jaker was fairly typical of a patrol base in the green zone (the well-irrigated, two- or three-kilometre-wide agricultural stretch either side of the Helmand River). A commandeered compound, it was reinforced by the ubiquitous wire-and-canvas collapsible HESCO Bastion containers filled with sand and rubble as quickly deployed protection; rudimentary but effective. Inside, conditions were sparse, unyielding and dusty: a tough environment that the human inhabitants shared with an eclectic population of mostly verminous wildlife, which was often potentially lethal. Food was always out of a tin or packet, water was bottled and beds were basic, uncomfortable, public and prey to mosquitoes. The ablution arrangements were also grim: showers, available on the rarest of days, were taken from a black bag that had been left to warm slowly in the winter sun, while a trip to the lavatories was always enlivened by the thousands of flies. Overhead cover against deadly munitions was also minimal.

Before leaving their base near Chepstow, south Wales, 1 Rifle's C Company had been reduced in size to just forty-four riflemen in order to be officer- and senior non-commissioned officer-heavy for the operational mentoring of the ANA's officers, non-commissioned officers and soldiers.

Now, on this March day in 2009, the men – and one woman – of Patrol Base Jaker were preparing for another 'advance to combat'. It was the second day of a three-day operation called *Operation Tor Tapus 2* – a contribution to the wider operation of *Operation Aabi Toorah 2C* – to which everyone was looking forward, albeit with the usual mix of anticipation, excitement and a modicum of well-controlled fear.

Despite a fatality a few days earlier, all wanted to help fulfil Major Watkins's desire to push back the enemy in Nawa in order to keep Patrol Base Jaker viable. This enduring role was in accord with the *kandak*'s longer-term aims of ensuring the continuing security of routes into the notoriously dangerous Nad-e Ali district that had, over Christmas 2008, been secured by Battle Group

Centre South during *Operation Sond Chara*. During the months following *Sond Chara*, the 4th *Kandak* had fought largely in support of the continuing series of *Aabi Toorah* operations with a secondary aim, in the immediate vicinity of Jaker, of setting the conditions for a strike against known enemy positions a few kilometres to the north of the patrol base.

With the day of the planned strike approaching, the need for positive identification of enemy bases and improvised explosive device (IED) factories was becoming more pressing, with the pressure mounting on the 4th *Kandak*'s commanding officer to provide this intelligence.

The Afghan colonel had been first trained by the Russians, then by the US Army's Fort Leavenworth Staff College. This whisky-appreciating, hardened veteran was being mentored, with his men, by Major Watkins and C Company. However, for the day's battle on 12 March – as was often the case – it was his operations officer, Major Hezbollah, who would lead half the *kandak* on patrol; the second half of the battalion was spread across a number of Battle Group South's other patrol bases.

The men of Battle Group South were expecting trouble; indeed they wanted it so that the intelligence on the location of enemy bases and IED factories could be obtained, and an engagement initiated. Given the softly-softly approach of the strictly enforced rules of engagement whereby ISAF are allowed to fire only if fired upon first, or if an enemy can be positively identified carrying a weapon or in the act of setting an IED, often ISAF and ANA forces would not come into contact with the Taliban unless they were able to tempt them into carrying out an ambush, or firing first.

However, when – if – trouble occurred, the battle group did not expect their task to be easy. Across the flat fields of the green zone, the Taliban enjoyed most of the advantages of surprise: they could afford to wait until the ISAF troops were in open ground and far away from good enough cover before opening fire: the chillingly entitled 'killing ground' could be of their choosing. They also had, if they chose, the advantage of climbing on to compound roofs for added height, which the ISAF troops could not do until they had captured a suitable building.

The rehearsal of concept for the second day of *Operation Tor Tapus 2* was played out in the dry earth within the patrol base's reinforced walls. A large-scale model of tracks and paths, objectives, lines of cover and potential emergency helicopter landing sites was marked on the ground using whatever suitable domestic object or building material was available: square-sided tins of processed meat often represented similarly shaped compounds, hedgerows looked remarkably like lines of green, army-issue socks, while the all-important map gridlines would be accurately reproduced with string drawn tightly across the huge models.

At the briefing's close, all knew their routes towards the area of operations. These were, by and large, north along the green zone; often very difficult to travel through due to its myriad drainage and irrigation ditches, walled compounds and semi-cultivated fields. Everyone knew the action to be taken on contact, and the positions of the emergency RVs should they become split from the main force as well as the casualty evacuation routes.

Major Watkins had delivered his main orders the night before, and the Company second in command, Captain Foster, had then amplified the logistic and administrative details that would ensure, among a vast array of other details, that the correct ammunition had been issued and in the correct quantities.

Nothing is more personal that a leader's orders, at every level. On them depend the success of an operation; by them men will live or die. Good orders can, occasionally, lead to defeat, but bad orders seldom lead to victory. Through his orders, a leader imposes his will on his subordinates; through his personality, a leader inspires confidence in his junior commanders and, through them, to every individual taking part.

The CO's orders were followed down the command chain by the company commanders and then by the troop officers, until finally, it was the turn of the section corporals to gather the men – with whom they live, fight and relax – around the rehearsal of concept model and 'walk the course', answering any questions that might arise, and testing comprehension of the orders.

In as close detail as possible from the map laid out in front of them, the men of C Company 1 Rifles understood the ground they would be fighting across during the next hours. They knew that it contained excellent routes for 'friendly' movement for, although scored by ditches, it was generally open, and offered good all-round observation; but it was also equally excellent country for an enemy lying in wait. This circumstance – common to warfare in Afghanistan – often meant that contacts were heralded by the cry of 'Advance to ambush', rather than the more usual 'Advance to combat'.

The time just prior to going into battle is always one of mixed emotions: the troop sergeants tease and cajole their section corporals, while the marines themselves exchange the kind of banter that only a serviceman about to face the enemy understands. They all know that the training, training and more training they have undergone will guarantee that most of the difficulties they encounter will be overcome, while the two D words – danger and death – are seldom uttered. Although wishing each other good luck, they all know that it will be their professionalism, fitness, superior fighting skills and field craft that will see them through. Luck is a bonus.

Under the increasingly knowledgeable eyes of their own junior commanders, the Afghan warriors of the 4th *Kandak* and their experienced mentors of C Company saddled up, after conducting their last-minute inspections:

'Weapons?' '*Check!*'
'Ammunition?' '*Check!*'
'Body armour?' '*Check!*'
'ID discs with blood group?' '*Check!*'
'Helmet?' '*Check!*'
'Goggles?' '*Check!*'
'Knee pads?' '*Check!*'
'First field dressings?' '*Check!*'
'Tourniquet?' '*Check!*'
'Morphine?' '*Check!*'
'Flares?' '*Check!*'
'Maps?' '*Check!*'
'Aerial photographs?' '*Check!*'
'GPS tuned in?' '*Check!*'
'Water? '*Check!*'
'Communications netted in, loud and clear?' '*Check!*'
'Spare batteries?' '*Check!*'

... followed by informal and very personal checks for the British members of the operation':

Fags and lighter? *Check.*
Photograph of partner/children/mother? *Check.*
Latest letter from home? *Check.*
Hip flask? Against the rules. Don't let the Boss see. *Check.*

'Right. Let's go!'

Leaving Jaker was never straightforward, and certainly never safe. Thanks to the regular ICOM intercepts of Taliban chatter, everyone was conscious that each route leading to and from the patrol base was under observation day and night, so every entry and exit had to be conducted in a manner that kept the 'dickers' – the Taliban's paid informers – guessing about the chosen destination.

The first objective lay just three and a half kilometres to the north, and the orders were to move out, quickly, in vehicles. C Company mounted their light, troop-carrying Pinzgauer trucks and WMIKs, while the 4th *Kandak* embarked in their own Ranger vehicles.

This time, the approach towards the suspect compounds was a deception, leading along a small canal and circumventing many small drainage and irrigation ditches, as well as walls and hedgerows. Either side of the rough, dusty track they travelled along, the stubble fields that had held the year's maize crop stretched away towards yet more mud walls and sparse hedgerows. This all-round, open ground was very definitely the two-edged weapon the men knew it would be: the troops of the 4th *Kandak* and their monitors could see, but they knew that meant they could be seen.

To the east lay a minor canal that was narrow, with slow-moving water. Beyond the open ground to the left, towards the true objective, the forbidding presence of featureless, beige-coloured walls – each one capable of harbouring an enemy – dominated the scene. Everyone knew that the Taliban fighters were watching them through their 'murder holes' in these walls. Everyone knew that each insurgent was waiting, ready for their opponents to reach the pre-chosen killing ground.

After painstakingly slow and careful progress, with men out in front conducting *Operation Barma*, the vital search – electronically and visually – for signs of IEDs, the order was given for the small convoy to halt and the troops to de-bus into their pre-arranged, all-round defence positions. Heavily laden men leapt to the ground and zigzagged towards ditches and hedges before flinging themselves on to their knee and elbow pads, keen to offer as low a silhouette as possible. The familiar cry of, 'On your belt buckles, lads!' went rapidly round the area.

Waiting by the canal, the three large multiples of the convoy, plus the Tactical HQ with the Afghan second-in command and Major Watkins, now prepared themselves for instant combat. Within each multiple, seven or eight men of 1 Rifles were monitoring and supporting about twenty ANA warriors. Although this was less than Company strength, it made for a potent individual force.

Each of the soldiers lying in the prone position in open ground was tense as they waited for the order to move. Their suspense was heightened by the ICOM radio chatter that was going on between the Taliban commanders and their men: chatter that indicated that the enemy knew where the *kandak* was headed and the route it was likely to take. It was a poor position to be in, but not unusual.

When he was ready, the Afghan major put his orders into action, giving the signal for the leading call sign to head off in a north-westerly direction, before it would swing south-west: all part of the continuing deception.

The objective that needed clearing of Taliban – if they were in residence – was a suspicious series of compounds about one and a half kilometres to the west. The ground was cut by interlinking irrigation ditches: some contained irrigation water but a few were used for domestic drainage – far less pleasant to wade or crawl through. Discarded stalks from the maize crop lay in uneven piles, providing no cover should a soldier need it but offering an effective trip hazard for men zigzagging at speed and under fire. The warm sun struggled through a dusty haze.

A short time to get into position and for some last-minute checks, and then, 'Come on lads! Advance to ambush!'

But events were not to be so cut and dried.

The leading multiple of the ANA began manoeuvring forwards by pepper-potting – individual fire and movement – ahead and slightly to the left of the HQ.

Behind, two further multiples fanned out either side, a tactical bound apart: all according to the rehearsal of concept so carefully laid out in the sand of Patrol Base Jaker. Under command of an ANA captain, the leading element now crossed the open ground without incident, reaching broken earth and rubble beyond. They were now just short of the nearest target compounds.

Suddenly, too suddenly as the rear multiples had hardly begun their own move forward into a tactical formation to give covering fire, the convoy's Tac HQ came under a long, sustained and accurate burst of small-arms fire from Taliban positions 100 metres to their left. Men jumped into ditches, their eyes, now at ground level, anxiously peering from beneath their helmets as they tried to identify the firing points. Then, silence.

As always, it was the silence that was as jarring on the nerves as the initial noise. But then a second burst of rounds slammed into the ground in front of the ditch, hurling dust and dry mud into the men's eyes and noses.

Quick orders were issued from the HQ: 'Half left! One hundred! Walled compound. Open fire!' There was no need for the orders as most lads knew where the firing was coming from, but they were given all the same. Another burst from the north-west riddled the HQ; then another from the west, pinning down the *kandak*'s command team. Unable to move, they were impotent for a few vital seconds.

And then a shout across the open ground, filling a new silence: 'Man down!'

This is the most chilling of all battlefield cries, even more terrifying than the sound of the danger-close crack of a Taliban 7.62 past the head.

The phrase is the cue to set in train a well-practised series of events. Heard not just by the men in action but also, via the battalion command net, by those at Camp Bastion, it also alerts the on-call MERT Chinook: the medical emergency response team helicopter.

The one person who perhaps had most to fear from this shout was Jaker's medic, Kate Nesbitt. Five foot three inches tall, her long blonde hair tied in a bun beneath her helmet, and serving with the OMLT to gain experience before being moved to more active, kinetic operations, she might have wondered just how much more kinetic it would get. But right then there was no time for analysing her situation other than to try and guess the, 'Who, where, how?' of her training, which raced through her brain on hearing the yell.

Kate was with the lead unit that had already crossed the most exposed area, and knew that no one with her was hurt. The casualty had to be somewhere back across the treacherous 100 yards she had just safely traversed. But she also knew there was no point in doing anything until summoned: there were three other subunits spread over the area, any one of which could have the 'Man down'. A dash in the wrong direction and she could be exposed unnecessarily. 'Wait out!' came the order over her radio.

Kate continued to listen to her personal role radio, but there was still confusion: all she could hear was, 'Man down! Man down!'

Then Major Watkins's calm voice came over the air: 'Amber Four Zero needs a medic. Now.'

Amber Four Zero was the major's own team.

'OK,' Kate thought, 'I know where they are.' Without a moment of hesitation, she leapt out of her ditch, keeping as low as she could as she went. She could see the major's helmet just poking above the parapet of his chosen cover, and ran towards it.

With her medical Bergen strapped on top of her Camelback water carrier, and that fastened on top of her bulky Osprey body armour, Kate looked for all the world like a small, fat tortoise as, crouching, she dashed to the major's ditch. Nobody who saw that zigzagging and stumbling sprint believed Kate would survive it. AK-47 rounds kicked up vicious, vertical spurts of earth all round her, following and preceding her erratic progress. But, seconds later and miraculously unscathed, she hurled herself into the bottom of the HQ ditch.

Amazed at her audacity, the men around her were still staring in disbelief at the direction from which she had come.

'Who, sir? Where?' she panted at the major.

'Over there.' Her company commander pointed across another seventy yards of exposed ground, which was being raked by enemy fire. 'Lance Corporal List. With John Foster and a couple of others.' Kate could just make out the helmets of two soldiers bent over a third.

Once again, Kate unquestioningly hauled herself out of the comparative safety of the shallow trench and began yet another dash though a hail of rounds. Although he had given the order, Andy Watkins, not quite believing what he was seeing, held his breath as he watched her run, praying that the covering fire he had ordered would do its job, and that he had not sent her to her death.

Being hit – dying – was not on Kate's mind as she hurtled towards the small group. From across the bumpy ground she could see Marine Mark Durry holding Jon List's head in his elbow. There was no sign of blood anywhere else: it had to be a head shot.

It was. A swift diagnosis was needed as she leapt into the small trench. From her first glance she knew instantly that it was going to be touch and go. The round appeared to have hit Jon's right shoulder webbing, ricocheting through the top of his left lip, through his teeth and tongue and exiting through his neck, blowing away much of his left jaw and the back of his tongue as it did so.

The danger lay with Jon's inability to breathe. As Kate knelt by his head, a new burst of fire smacked into the ground around their position.

'Get down, Kate. On your bloody belt buckle.' Captain Foster tugged at her webbing as, kneeling, Kate struggled to rid herself of her medical Bergen. She couldn't do that lying down.

Jon List was barely alive: drifting in and out of consciousness, spluttering on his own blood and with a failing breath rate – nor were they proper breaths, either. Kate tried desperately to find a pulse.

While she continued her search for vital signs of life, the situation was being relayed back, continually, to Camp Bastion, as well as being recorded in the Brigade HQ's log book in Lashkar Gah:

15.41   Reference contact Nawa. One casualty. Gunshot wound to neck.

15.45   Do you require attack helicopter to launch ASAP?

15.46   Yes. Call sign still in contact and pinned down. Emergency helicopter landing site is set at Grid XYZ.

With no sign of a pulse and only those bloody, erupting gasps for breath, Kate knew she had to work fast to get air and fluids into Jon List's dying body. Air first. Once she had inserted a nasal tube through a nostril and past the back of his damaged throat, she then positioned him on his side to prevent his body fluids flowing down the back of his throat. She was then able to insert an IV tube into a vein and force 250mils of saline solution into her patient. The response was an almost immediate weak but workable pulse rate.

With Jon's pulse rate improving and no longer struggling for air, he started to come round: he was not properly conscious yet, but alert enough to feel pain. A new worry struck Kate: supposing he had brain damage from splinters inside the back of his mouth? She checked his pupils. Normal. With no further adverse signs, and confident that he was responding to all of her demands, Kate jabbed a phial of morphine into his leg.

The nasal tube was narrow and blocked easily, and in his shocked state, Jon started fighting it. 'Jon,' Kate said harshly into his right ear, 'it'll help. Don't try to breathe through your mouth. I want to be really chuffed with you. Let the tube do the work.'

If it didn't do the work, Kate was mentally preparing herself to carry out a tracheotomy. But she reckoned that it wasn't needed yet; Jon was fine as he was, and any more trauma would have been a mistake. (The operation was to come later when Jon was back in the UK. In the controlled environment of Camp Bastion they managed to stabilise him by packing everything down and placing a breathing tube down Jon's throat.)

With her patient stabilised and out of pain, Kate's thoughts turned to the next demanding task: casualty evacuation. Neither Jon nor the team protecting and saving him were yet out of danger. Taliban fire was still kicking up the dust and although – thanks to the continuing fire fight around them – this was becoming more desultory and less accurate, she was all too aware that it would take only one stroke of bad luck …

*

The moment the 'Man down' cry had been received in Camp Bastion, the medical emergency Chinook had been alerted. Now Andy Watkins needed to choose a safe landing site for it, and quickly. He contacted Kate on her PRR, telling her that there would be a delay while his team chose a safer landing site than the one chosen earlier, and it was Barma-ed. She simply had to keep going.

While this was going on, Captain Foster relayed the NATO standard nine-line medical evacuation request, based on a series of quickly transmitted and crucial facts.

Kate kept going, monitoring Jon by constantly checking that his pulse rate was acceptable and that his breathing was satisfactory. Continually talking to herself, she kept up her life-saving routine: 'Breathing? Fine.' 'Rate of fluid input? Fine.' 'Check the nasal tube is clear of clotted blood. Done.' But each time she removed the tube, there was a very real risk that Jon would choke before she had time to clear the blood and mucous from it and reinsert it back past his damaged throat.

She was winning, but for how long?

Ten minutes after the initial contact, authority was given for a series of attack helicopter strikes on the area, which helped calm the situation and allow Amber Four One, the reserve team, to dash across the field and prepare Jon for the move to a large field that the major had deemed the best place for the helicopter to land. The team arrived quickly, the Afghan interpreter carrying and then preparing the field stretcher while the others helped Kate roll Jon on to it.

As they removed Jon's constricting body armour and helmet, Kate could suddenly detect no sign of his tongue. Concerned that he was sucking it down his throat with each inward breath, she ordered that he remained strapped on his side on the stretcher.

As the team began making its tortuous way to the emergency landing site, Kate stayed by Jon's head. The route was rough and at two stages the stretcher had to be manhandled across small waterways that were as deep as Kate was tall. Each time they reached the far side, Kate checked Jon again and at one point injected a second dose of morphine. Jon could not speak but he could squeeze Kate's hand in response to her simple questions.

At last, after one very long and tiring kilometre over exposed ground, the team reached the landing site. At their approach, the in-coming Chinook flared out, dipping its tail to reduce height and speed rapidly, prior to landing, before – now almost level and stationary – disappearing into the customary brown-out of dust and small stones. The stretcher party quickly covered Jon's mouth and nose: choking dirt could have killed him as instantly as a blocked nasal tube.

With the dust settled enough to see, the on-board emergency team leapt off the stern ramp. Kate ran forwards and swiftly gave them a verbal handover of information on Jon. As she watched him being lifted into the helicopter forty-five minutes after he had been wounded, gingerly but with haste, she was confident that his prognosis was good.

At that moment Kate, of course, had no idea that Jon List would not wake up until six days later, when he was in Selly Oak hospital, Birmingham; nor that, after the fitting of a plate and a few false teeth, he would recover fully, keen to get back to work.

Nor would she have known that she would become only the second female recipient in the history of the Military Cross for her actions, which were summed up by the official citation: '*Nesbitt's actions throughout a series of offensive operations were exemplary; under fire and under pressure her commitment and courage were inspirational and made the difference between life and death. She performed in the highest traditions of her Service.*'

CHAPTER ONE

MISSION

Searing heat, numbing cold, flash floods, glutinous mud, soft sand, all-enveloping dust storms, mosquitoes by the million, flies by the billion, fast-scuttling camel spiders, lethal fat-tailed scorpions, deadly saw-scaled vipers and a seemingly endless supply of Taliban fighters.

Welcome to Helmand Province, southern Afghanistan: the most dangerous place on earth. To many of the sailors, marines and soldiers of the 3rd Commando Brigade arriving in September 2008, none of this was new.

The first time 3 Commando Brigade had deployed to Afghanistan was in 2003 for *Operation Jacana*. Then, for *Operation Herrick 5* between September 2006 and April 2007, the brigade relieved 16 Air Assault Brigade and the stalwart 'Toms' of 3 Parachute Regiment, whose tussle with the Taliban during *Herrick 4* had been noble but bloody.

For *Herrick 4*, Brigadier Ed Butler's 16 Air Assault Brigade had included one Parachute Regiment battalion – a manoeuvre unit supported by cavalry, sappers and gunners – while the embryonic operational mentoring liaison teams were supplied by 7 Parachute Regiment and the Royal Horse Artillery.

The remit in those earlier days had been to defend static locations – the infamous platoon houses – as Butler's brigade did not possess vehicles (such as the Viking) suitable for the campaign of manoeuvre he would rather have been carrying out. Added to this, the unforeseen strength of retaliation from the Taliban necessitated putting the concerns of the operational mentoring and liaison teams to one side for a time; all hands were needed in the face of unexpectedly fierce Taliban opposition.

*

From the first, *Herrick 5*, with its more aggressive terms of reference, threw up a number of operational shortfalls in the constitution of the Commando Brigade, experience that was to prove useful in subsequent Herrick tours. Although it deployed with two commandos – 42 and 45 – the latter Commando was split into OMLT teams across the province, thus leaving the deployed brigade with just one manoeuvre unit. As an emergency measure, the then Brigade Commander, Brigadier Jerry Thomas, managed to conjure up what was, in effect, a second Commando out of the Command Support Group at Brigade HQ. The new manoeuvre unit, known as the Information Exploitation or IX Group, performed valiantly alongside 42 Commando, but it was constrained by limited command and control elements, as well as denying the Brigade HQ its own command support function.

The lesson, however, was quickly learnt: three manoeuvre units deployed on *Herrick 6*, one of which supplied the OMLTs; four deployed on *Herrick 7* with, again, one supplying the OMLTs; five battalions deployed on *Herrick 8*; and then it was back to four for *Herrick 9*, although the Helmand reserve battalion was taken under command for specific operations.

Under normal circumstances, the 3rd Commando Brigade should deploy with three commandos: 40, 42 and 45 (the Brigade did not deploy with 40 Commando for *Herrick 9*), with 1 Rifles as a fourth manoeuvre unit. Supporting arms within the Brigade's permanent order of battle are represented by the Commando Logistic Regiment; 29 Commando Regiment, Royal Artillery; and 24 Commando Regiment, Royal Engineers; plus smaller units such as the Command Support Group which, on operations, is often re-configured into the Information Exploitation Group, as happened for *Herrick 5*. The Brigade Reconnaissance Force is a squadron within the Command Support Group (now renamed

30 Commando Information Exploitation Group Royal Marines) while 539 Assault Squadron is another, smaller, supporting maritime unit within the formation.

For specific operations, other organisations can be attached to Commandos: during the Falklands campaign, for example, two troops from B Squadron of the Blues and Royals served with distinction; at Al-Faw in Iraq, two troops from C Squadron the Queen's Dragoon Guards assisted the assault; C Squadron, the Light Dragoons, were attached to the Commando Brigade for *Herrick 5*; while for *Herrick 9* it was, again, the Queen's Dragoon Guards, while there was also support from Harriers from the Naval Strike Wing. Again, many smaller but no less vital elements of the British armed forces can be attached, too, such as the Royal Engineers' Post and Courier Unit, and elements from the Royal Navy's three rotary-wing naval air (Commando) squadrons 845, 846 and 847.

This cross-services support emphasises strongly that Afghanistan is not just the army's war. With all these personnel, plus those in the Royal Naval surgical support teams (not forgetting the Royal Marines Band's service who, when on operations, act as medical orderlies and stretcher bearers, with their role of entertaining the troops only fitted in as other tasks allow) at Camp Bastion's field hospital, the number of men and women from the naval service, for example – as was the case during the Brigade's earlier deployment in Afghanistan when it was about 55 per cent – was nearly 50 per cent of all British service personnel in theatre.

Good equipment and vehicles are also crucial in warfare, the latter proving to be particularly controversial throughout ISAF troops' time in Afghanistan.

Throughout much of *Herrick 5*, the Commando Brigade enjoyed the unarguable advantage of the armoured Viking, a tracked, amphibious vehicle that provided the up-to-then-missing but crucial component for modern, rural, counter-insurgency operations: rapid manoeuvre. Through the intelligent use of these all-terrain, armoured vehicles, in addition to the open, un-armoured but heavily armed WMIKs, the commandos were able to bite away at the Taliban, sapping their military strength, if not their will.

During *Herrick 5*, the new-to-battle Vikings had supplied not only protection from small arms and the ubiquitous rocket-propelled grenades (RPGs) but had offered swift manoeuvre across the trackless *dasht* or desert – especially in those places impassable to wheeled vehicles. They could, too, ford the Helmand River when needed. But they are not perfect – no vehicle in battle ever has been. The Viking's biggest threats came from the thousands of legacy mines laid by the Russians – often re-laid by the Taliban as IEDs – and the deadly home-made IEDs themselves. Marines were wounded and vehicles lost during the Brigade's first deployment on *Herrick 5* due to this uncollected – and reused – ordnance left over from the Russian occupation of Afghanistan.

Despite adverse media comments about men being forced to travel in soft-skinned, open vehicles, many marines preferred to approach a battle with continual situational awareness, and this battlefield requirement is not one easy to gain, nor keep up to date, from inside the confines of a Viking. Many Royal Marines opted for travel in the open WMIK Land Rovers, with yet further marines preferring to travel in the more lightly protected 'Snatch' Land Rovers, as the extra height provided by these vehicles is invaluable for peering over the miles of hedges and compound walls. Senior commanders during *Herrick 9* often found they had to justify the continued use of all twenty-seven on the Brigade's strength, as they were loath to lose them. At the same time, it was noted how armour-friendly marines could become when bullets and rockets began whizzing about their ears ...

However, the standard set by the Commando Brigade's *modus operandi* during *Herrick 5* – they relied on greater manoeuvrability and had a more cohesive personnel structure for most knew each other and had often served together – became almost too successful. During the three six-month tours between April 2007 and September 2008 – *Herrick*s 6, 7 and 8 – army brigades whittled away at the Taliban so successfully using this approach that, realising they would almost always be defeated in fixed battles, the enemy began to rely more and more on asymmetric warfare.

Forced by their failures in the field, the Taliban began to increase their expertise in, and use of, IEDs, while also starting to employ suicide bombers to deliver some of them. From the end of *Herrick 5* to the beginning of *Herrick 9*, the incidents of IEDs increased from roughly two a week to nearly forty. As a counterpoint, the find and turn-in rate was to increase dramatically and, while the IEDs were certainly an irritation as each one had to be marked and guarded before it could be dealt with, it was also an indication that the civilians, fed up with suffering their own casualties, were passing on information more readily than before.

However, the increase in IEDs also meant that Vikings and even harder-skinned vehicles were now becoming death traps rather than safe havens, and the increase in suicide bombers, command-detonated improvised explosive devices, road-side bombs, legacy mines and all manner of improvised devices were to form the largest threats to life during *Herrick 9*.

In response to this, and with the Commando Brigade's deployment on *Herrick 9*, it was to be the turn of the medium and heavy lift helicopters – the Sea Kings, Chinooks and the American Sea Stallions – to help take the war to the enemy, and in a way they had seldom done before in Helmand. The aircraft that were in Helmand during the winter of 2008–09 were more usually earmarked for the resupply of men and logistics and for casualty evacuation, but their use for night-time commando attacks – a prime *modus operandi* of the Royal Marines – was to be an important feature during the Brigade's second tour.

Another improvement in battlefield mobility by the time of *Herrick 9* was the Jackal armoured patrol vehicle. Those who favoured the ubiquitous WMIK for patrolling and fire support quickly changed their allegiances as this armoured, open and heavily armed vehicle gave much better protection to its occupants in a mine strike.

Good equipment is all very well, but most marines would agree that it is their training that makes them the effective force they are. Before its second deployment to Helmand in the autumn of 2008, it was decided that the Commando Brigade should renew its day-job qualifications. In early 2008, it set about honing the primary skills of amphibious, winter and expeditionary warfare in Norway.

On their return from Norway in spring 2008, the component parts of the Commando Brigade entered their more specific *Herrick 9* pre-deployment training, which was largely copied from 16 Air Assault Brigade's preparations for *Herrick 8*. The soon-to-be-deployed Commando Brigade liaised a great deal with the on-the-ground AAB, neither team adhering too closely to the advice from the Permanent Joint Headquarters as, following their experiences in *Herricks 4* and *5*, gained experience within both Brigade HQs indicated what was necessary. As a result, the Commando Brigade drew up a complex and relevant training programme that would take it from April 2008 through to that September, including an early package of pre-deployment training that allowed some of the young marines from 42 and 45 Commandos to attend driver and language courses. One lesson learned from *Herricks 4* and *5* was that they couldn't start training for the fundamentals too soon.

Tragically, however, early preparations for the tour's vital mission rehearsal exercise that was held in July 2008 were struck by tragedy: in June, a corporal was killed in a freak accident involving his and another vehicle, and then in July, a marine was killed when the Land Rover in which he was a passenger overturned. 'Hard training for an easy war' is a well-worn adage, but this was a tragic start for the Brigade and in particular for 42 Commando, where the two men had been serving.

A third, thankfully non-fatal accident was to occur which, while it did not have the same devastating effect as the two deaths had on 42, was to have a profound impact on Brigade HQ.

In April 2008, Brigadier F. H. R. Howes became the Brigade Commander and, as such, was destined to accept responsibility for Task Force Helmand from Brigadier Mark Carleton-Smith OBE, who was then commanding 16 Air Assault Brigade.

Another change of key personalities had also taken place on the Brigade's return from *Herrick 5* when Lieutenant Colonel Al Litster had taken over as Chief of Staff. His new responsibilities were to be tested to the full when, in August 2008 and one month before the Brigade's departure for Helmand, he received a telephone call informing him that his brigadier had hurt himself badly – almost critically – in an accident. In a freak water-skiing 'wipe-out', Howes had broken his pelvis, a fracture that would prevent him from continuing in command.

This was a blow not only for the Brigadier but in particular for Howes's teams, some of whom – in particular the Royal Engineers – had already deployed to Helmand and were operating in accordance with his preliminary orders.

The Deputy Brigade Commander, Colonel Martin Smith; the Chief of Staff, Lieutenant Colonel Al Litster; and the Deputy

Chief of Staff, Lieutenant Colonel Rory Copinger-Symes, faced a problem. The Brigade faced a problem. In Brigadier Howes's absence, Colonel Smith called the commanding officers to a conference at the Brigade's Headquarters in Stonehouse Barracks, Plymouth.

Colonel Martin Smith was commissioned in the Royal Marines in May 1984 and, via the usual appointments and specialist courses, plus parachute training and the inevitable Arctic deployments, had enjoyed a tour as the second in command of 45 Commando. During *Operation Telic 1* in Iraq in March 2003, he had been seconded to the United States Forces Headquarters for what they called *Operation Iraqi Freedom*. Eventually, he had commanded the Command Support Group and had helped establish the first ever Information Exploitation Group. He had joined the Brigade shortly before *Herrick 9*, not expecting to command except when Buster Howes was on leave. He had planned to accompany the brigadier on his pre-tour recces, then remain in Afghanistan to manage the handover between 16 AAB and 3 Commando Brigade; it was considered necessary to place a colonel on the ground early, to inject continuity at senior level.

Following Howes's injury, and as Deputy Brigade Commander, it was Smith's job to step into the commander's shoes for as long as it took the system to appoint a new one, or until it accepted the deputy as a permanent fixture. Although it was a personal tragedy for Buster Howes, his accident would mean that another officer would be in line to accept one of the more unusual and swift 'pier-head' jumps that happen occasionally.

In early September 2008, Smith visited London for briefings before flying to Pakistan and Afghanistan, by way of an initial recce. The Brigade's training in the United Kingdom was in full swing with some units already deployed, while the remaining commanding officers were still arguing for the last yard of manpower poss-ible. These were matters with which the staff was wrestling and Smith had to leave them to it.

At this first meeting without Howes, Smith needed to satisfy himself that every CO around the table was content with the initial operation order for Helmand. Unsurprisingly, they were.

Smith knew his way around the Brigade as well as anyone and, at first, thought that he might have been asked to replace Howes, especially as he had successfully conducted what should have been his brigadier's initial visit to Afghanistan and Pakistan in September. But the Corps – understandably, and accepted by all concerned – had other plans and, on 15 September, appointed Brigadier G. K. Messenger as the new Brigade Commander.

Transfer of authority – the handover from 16 Air Assault Brigade to 3 Commando Brigade – in Helmand Province was just three weeks away.

Gordon Messenger was a mountain and Arctic warfare specialist for whom drinking real ale was one of life's pleasures. As the Chief of Staff (formerly the Brigade Major), Messenger had served with the Brigade in Kosovo, for which he had been appointed an OBE. Later, he had commanded 40 Commando during an operational tour in Afghanistan. *Operation Telic 1* in Iraq had then followed, when he had led 40 Commando in the initial assault on to the Al-Faw Peninsula preceding an attack against Iraqi armoured forces at Abu al Khasib, for which he was awarded the DSO. Having joined Joint Force Headquarters as Chief of Staff in July 2004, more Middle East appointments cropped up, which involved him in *Operation Garron*, the 2004 tsunami relief operation; and *Operation Highbrow*, the operation that came about in July 2006 following intense fighting between Israel and the Hezbollah that had triggered a major evacuation of British nationals from the Lebanon. More to the point, Messenger had enjoyed six months in command of the *Herrick* preliminary operations study where he, more than any other, had been instrumental in drawing up the initial plans for the continuing series of operations. Finally, between operational appointments, the new Brigade Commander had also attended the Canadian Forces Staff College and the United Kingdom's Higher Command and Staff Course. It was difficult to imagine an officer more qualified than he to command the Brigade.

When Messenger was appointed, Smith was already in theatre and was able to conduct him on his pre-deployment recces; recces that included meetings with 16 Air Assault Brigade's Brigadier Carleton-Smith and Helmand's Provincial Governor, Gulab Mangal, in his capital, Lashkar Gah. Messenger then returned to Plymouth, leaving Martin Smith to provide the all-important continuity.

Through the summer of 2008, the Commando Brigade staff had been watching what had been happening in Helmand with horror and sympathy, in particular the logistical and tactical nightmare of the delivery of a turbine to the barely operational Kajaki dam during *Herrick 5*.

Brigadier Carleton-Smith had wanted to concentrate on the tricky district of Marjah, but 16 AAB, through the decision of others, had had to commit fully to the Upper Sangin Valley and to the Kajaki project.

Aided by Commando Brigade's actions, which managed to capture the last heights that dominated the area, thus preventing the Taliban from exercising any control over the dam, 16 AAB had done an excellent job across their tactical area of responsibility.

Yet, despite having to concentrate so much attention on the Kajaki project, the AAB's Chief of Staff had still kept his situational awareness of other areas. He was happy that he was up-to-date with events in and immediately around Lashkar Gah, but was more concerned with the green zone of Marjah district to the south-west of Lash. There were reports that the area was becoming ungovernable, with rumours that ANP policemen had been seen running away rather than choosing to take on the Taliban.

Herrick 9's answer to this was to be *Operation Sond Chara*, viewed as the preliminary clearance of most of Nad-e Ali and parts of Marjah, to pave the way for their successors – 19 Light Brigade – to thrust north-east into the Babaji district during *Operation Panther's Claw* between June and August 2009. *Panther's Claw* had originally been conceived and planned by 3 Commando Brigade as *Sond Chara II*, but as the Brigade began to run out of time and assets, 19 Brigade's Chief of Staff had been flown in from the UK to assist with initial planning. Consequently, the operation was successfully executed by the Light Brigade during *Herrick 10*.

During *Herrick 7*, a temporary battle group had been formed to cover the Musa Qal'eh area in the north-east of the province, after the town had been retaken; then, in 2008, another battle group had been established in the south by the United States Marine Corps's (USMC) 24 Marine Expeditionary Unit, to cover Garmsir and the snakeshead stretch of the River Helmand. When the 24 MEU left during 16 AAB's tenure, the British reassumed responsibility for that area.

With the USMC's withdrawal from the south, they and the US army would enter an unwritten competition between themselves for ascendancy in Afghanistan, especially across the north-west corner of Helmand, the new American area of operations.

The withdrawal also left the British with a more manageable slice of the province. Britain's redefined areas of responsibility now started in the west with Marjah, whose western limit was marked by the town's substantial canal. This water course marked the edge of the whole area, and divided it from the vast expanse of desert to the west – an undulating, wide-open *dasht* of sand, gravel and scrub, defaced by rocky outcrops, steep-sided defiles and gorges. The desert stretched away for more than eighty kilometres towards Nimruz Province in the west, offering an area that the Commando Brigade would treat much as it would have done an amphibious operation's open sea flank: as a vast manoeuvre space. To the north-east, along the Helmand River, lay the green-zone sprawl of compounds that was Babaji then, further on, Gereshk with Camp Bastion in the desert to its west. Sangin was yet further north, then, eventually Kajaki with its dam that straddled the border with Kandahar Province. To the south, where the green zone narrowed to the more normal average width of less than five kilometres, was Garmsir district, then, south again, the snakeshead area and fish-hook curves of the river as it gently bowed south and then west.

Finally, the reallocation of areas of responsibility also allowed the Estonian military contingent, then based in Now Zad, to come under direct command of the Commando Brigade.

Other, subtler changes had also taken place. Brigadier Mark Carleton-Smith had felt that there would be a danger that the Air Assault Brigade might become blinded by the intensity of the fighting they had to undertake, and that they might fail to address the needs of the Afghani people. While that had been understandably true in earlier *Herrick*s, it was a problem he had chosen to address and 16 AAB's actions had laid a very solid foundation for 3 Commando to build on during *Herrick 9*.

A less subtle change was the considerable increase of personnel in, and the growing efficiency of, the Afghan National Army, to the point that when 3 Commando Brigade arrived in theatre, the ANA's strength alone was up to 55,000 due to successful recruiting and a raising of efficiency, and thus morale, by recent OMLTs.

Thanks to the speed of his appointment and the urgency of the situation, Brigadier Messenger had been obliged to take over Buster Howes's training programme and his initial operations order, although as part of Regional Command (South), these were largely pre-ordained through what had been put in place by 16 AAB.

A brigade commander has to make his own mark on such a fundamental section of his brigade's orders but, rather naturally, Messenger had decided to wait until he had been in Helmand long enough to make his own assessment. Senior staff officers were expecting this as they knew that there are always little differences between individual commanders' assessments when studying any given tasks. Happily, however, Brigadier Messenger was content with the majority of what had been done in advance.

Assessment made, part of the Brigade's new orders was to disrupt the enemy along the Helmand River and keep them under pressure. The main effort was to focus on the centre of the province, yet pressure still needed to be applied to the Taliban in the Upper Sangin Valley, and then as far south as force could, sen-sibly, be effective.

Throughout Helmand, the trend for troops in contact was creeping remorselessly upwards; no more so than during *Herrick 9* when the Commando Brigade had more thrown at it on a regular basis than in any other tour; and that did not include the IED count.

This increase in troops in contact initiated debates in Kabul, Washington and London, which helped to demonstrate why Helmand was – and remains – so important. Not everyone agreed with this position: in some quarters of NATO, the accusation was made that the British forces had a very Helmand-centric approach. However, 3 Commando Brigade's point, made in the middle of their tour, was that in February 2009, for instance, 39 per cent of everything that happened throughout the whole of Afghanistan happened in Helmand, suggesting that the enemy had decided that Helmand was to be where they concentrated their main effort, and the sooner NATO realised that the better.

The on-the-spot view was unequivocal. If incidents across the British areas of responsibility were added together, then they accounted for a high percentage of all that happened in Afghanistan, with only 24 per cent of the action happening in the American area of operations. However, it was a fine balancing act, as it was felt strongly that NATO should never come to see the situation as the British fighting in isolation; medical support, for instance, was a NATO-wide system, and the same was also true of helicopter support: often other nations' aircraft conducted casualty evacuation and ground support for British troops. British helicopters may have been thinner on the ground than the troops would have liked but, as this was a NATO-wide concern, when helicopters were needed, they were almost always available.

On 8 October, Brigadier Gordon Messenger arrived in Lashkar Gah to accept the transfer of authority for Task Force Helmand from Brigadier Mark Carleton-Smith. He was presented with a vibrant going concern following a successful and highly intelligent deployment, during which Carleton-Smith's brigade had kept the Taliban tide in ebb rather than flow.

Task Force Helmand was a British ground force, which at the time also contained a Danish battle group and an Estonian mechanised infantry company. Along with other, more military missions, it worked closely with Lashkar Gah Provincial Reconstruction

Team to help them achieve their aims, usually by providing protection for PRT projects.

As expected, the handover between the units of the two brigades during the first weeks of October and across the province was exemplary, thorough and much appreciated at all levels. 3 Commando Brigade would be able to hit the ground not just running but, as it was to turn out, at a positive gallop.

It was not until two days later that the Brigade Commander met his full staff, in a meeting that was to prove interesting in more ways than one. At that time, the incoming brigadier's intention had been to consolidate the gains made by 16 AAB and 24 Marine Expeditionary Unit in securing the focus districts of Lashkar Gah, Musah Qaleh, Sangin and Garmsir, and to concentrate all efforts on enabling the implementation of the Helmand 'road map' by the area's Provincial Reconstruction Team. Although early major combat operations were not anticipated, the possibility was real enough, especially as the outgoing brigade had warned that Provincial Governor Gulab Mangal's position was being undermined, and that the security situation in Marjah and Nad-e Ali was deteriorating.

It was time to take stock of the new Brigade's dispositions. Messenger's headquarters at Lashkar Gah were known simply as Lash PRT as they were co-located within the same cramped compound as the Helmand Provincial Reconstruction Team, headed by Hugh Powell of the Foreign and Commonwealth Office, with Colonel Haydn White, Royal Marines, as his deputy. White was also a member of the Brigade command team.

Governor Mangal's Provincial Governor's quarters were three kilometres away from Lash PRT, next to the river to the south-west. Powell, accompanied by White and often by Messenger, would travel to Mangal's compound every day, usually at speed and always in armoured vehicles for safety. Highly regarded by the British as a provincial governor with whom the Foreign Office representative and the commander of Task Force Helmand could do business, Gulab Mangal had survived thirteen assassination attempts since being trained by the Russians as one of their local commissars.

That Task Force Helmand had its headquarters in the provincial capital in the first place was thanks to Brigadier Jerry Thomas's actions during *Herrick 5*. Before that, Task Force Helmand's headquarters were, inexplicably, in Kandahar, to the east of, and outside, Helmand Province.

By moving his headquarters, Thomas had demonstrated to government departments, both Afghan and British, that he meant business. This repositioning had helped considerably in creating and maintaining the favourable conditions within which the PRT's outstations across the Brigade's area of responsibility had helped develop provincial government.

There had been, though, some opposition to the move, for Lash PRT had only one helicopter landing site and was in an area known for its suicide bombers. Added to this, the concern at the time had been that the increase in activity there would be visible to anyone studying the place, and a resulting increase in bombs would be likely. The Department for International Development and the Foreign Office had also worried that such a large HQ would entail a huge domestic, administrative and security upheaval. However, once established plans for a rebuild were in place, the go-ahead was given, and on 6 November 2006, Task Force Helmand began settling into its new and expanding home of 'Lash Vegas', as it was almost universally known.

With one and a half miles of perimeter walls, Lash Vegas was not too cramped, despite having to cater daily for anything between

500 and 700 people. (It was to become rather more full when home to as many as 1,500 men and women, as happened following the formation of Battle Group Centre South during *Herrick 9*.) The newly expanded compound contained permanent offices and accommodation blocks, which were smart, modern and air conditioned; inevitably, however, there was a hierarchy to the living accommodation: the more elevated ranks enjoyed the better accommodation, with the twelve most senior officers guaranteed en-suite rooms fitted with internet access. Further down the food chain, accommodation consisted of four- or five-bed dormitories, then, lower still, men slept in tents – but tents with air conditioning and wooden-framed camp beds. When the Estonians were in residence, they lived in their vehicles. Lash Vegas also contained two helicopter landing sites, although one was more usually used by the military transport department as a vehicle park: it was not possible to keep aircraft *in situ* there, either to refuel or re-arm them, as the landing sites were situated in a huge open space in the centre of the compound. Although the whole camp was contained within HESCO Bastion block walls and surrounded by an open 200-metre zone, nowhere was mortar-proof. Nor was the compound immune to 107mm rockets.

What was perhaps surprising was the realisation that almost all supplies, especially food, were brought from Kandahar in local 'jingly' lorries, owned and driven by Afghans, then, once stores were broken down for distribution to the forward operation bases – and particularly those along the Upper Sangin Valley – the trouble would begin. The 60-truck combat logistic patrols only had a choice of two routes between their destinations, and the potential for ambushes and IED strikes was high.

Dubbed the 'Viceroy of Helmandshire', Hugh Powell, a two-star diplomat – major-general equivalent – was responsible for the sixty-six civilian and fifty-five military members of Helmand PRT. Although a quaint sobriquet, it was an unhelpful tag as it bought into the 'Helmandshire' image that the Americans accused the British, wrongly, of fostering.

Powell led the effort to develop local government based on the Helmand road map, the agreed ethos that lies at the heart of all progress in Afghanistan, and the document that helped 3 Commando Brigade decide what it needed to achieve during *Herrick 9*. It was a complicated matrix based around eight simple tenets concerning: security; politics and reconciliation; governance; rule of law; counter-narcotics; strategic communications; socio-economic development; and district stabilisation.

The relationship between Task Force Helmand and the PRT was, on paper, simple: the International Security Assistance Forces provided the security space for Afghan government development, with a steadily increasing civilian input at all levels and across many disciplines.

To assist the PRT's work in the local district centre, a number of stabilisation advisers, or STABADs, were present across the province: in Sangin, for example, Lieutenant Colonel Jim Morris, the commanding officer of 45 Commando, worked closely with his civilian STABAD counterpart, Nick Pounds, a retired Royal Marines brigadier and pilot. As happened at Lashkar Gah, Morris and Pounds worked closely with the local governor of the area, Haji Faisal Haq.

It was vital that the two teams – military and political – worked hand-in-hand. With the Brigade's planning team based in the PRT's office, it ensured that the plans made for all the Brigade's military operations fulfilled the over-arching aims of bringing peace, while strengthening governance in all their areas of responsibility.

By far the most effective projects by which these aims could hope to be achieved was through the distribution of wheat seed and the establishment of a successful voter registration programme, while other civil military projects would come and go as the situation developed.

It was known that wheat seed distribution and voter registration were already part of the future Helmand battle rhythm; a fact that had been handed over with the transfer of authority. Before 3 Commando Brigade's arrival in September 2008, there had been a tendency to build things – schools, wells, roads, clinics – in order, all too often, to show that something was being done. Unfortunately, much of this meant little to the locals and it was proving that it simply was not possible to build a way to security in Helmand. Too little thought went into the end product of these projects: just because money was available for the bricks and mortar was no guarantee that there was anything to put into the resulting buildings. Who, for instance, would supply the teachers, doctors, salaries, books and medicine needed to run schools and clinics, or maintain the roads and wells? In short, there was often little logic and even less planning behind this so-called progress.

From the beginning of the Commando Brigade's tenure in Helmand, it was necessary to step back a pace, and attempt to link these projects to the appropriate Afghan ministry. 'Yes,' was the new message, 'we'll build your school but only if you supply two teachers and your Ministry of Education takes on the responsibility for salaries and equipment.' It was a new approach that would forge particularly strong links between the PRT and the Brigade.

This new approach also incorporated only undertaking projects that would endure, particularly as they were introduced as projects provided by the district or provincial governors, or even by President Karzai himself. These schemes included the building of a hard-topped road and bridges network – allied to a good level of security along the route – as they were considered enduring and more totemic than 'just another five wells'.

The PRTs also knew that developing a road network would help in the fight against narcotics. The green zone's fertile and superbly irrigated land was targeted by the United States Agency for International Development, who had instituted all manner of agricultural projects such as the growing of chillies, raisins and pomegranates, very successfully. But these goods still lacked the infrastructure to get them to a point of sale and, in desperation, the subsistence farmers of Afghanistan always returned to cultivating opium poppies, the staple that had provided them with a cash crop for many years. As the narcotic 'barons' in the area would then collect the raw heroin, paying for it there and then (if the farmer was lucky), this also negated the need for the farmer to go to market; itself a near-impossible exercise without proper roads and no transport. And, as most of the drug barons were now agents of the Taliban, it gave the Taliban an incredibly strong economic hold and influence over the area.

Helmand PRT would coordinate these new projects by asking the local Afghan elders where they wanted their new road or new school to be, and once that had been established, the Brigade staff would be brought into the discussions in order to check that the appropriate level and scale of security could then be provided.

Although the security situation was never mature enough for Task Force Helmand to achieve all that was asked of it, a remarkable amount was managed during *Herrick 9*. For the all-important (and extremely successful) voter registration to take place, a similar plan was drawn up by the PRT and made to work by Task Force Helmand. *Operation Janub Killi* was a subset of *Operation Sond Chara* and provided a security umbrella to cover the voter registration sites and offices – three in Garmsir, three in Nad-e Ali and two in Nawa, running parallel to the other reconstruction projects of the counter-narcotics operations and wheat seed distribution programme.

Heading the list of 3 Commando Brigade's major Brigade units in theatre was the Command Support Group (CSG) of 3rd Commando Brigade, commanded by Lieutenant Colonel Andrew McInerney. Known as the Information Exploration Group (or just the IX Group), and unique to the Royal Marines, the IX Group's mission was to win the information battle and thus shape the battle space through finding, understanding, influencing and delivering intelligence by using surveillance, target acquisition and reconnaissance (ISTAR) assets, combined with communications, information systems and information operations (including electronic warfare and psychological operations). The IX Group brought all these effects together under one single, coordinated tactical plan that included command of the BRF, and a UAV battery among other surveillance capabilities.

In the north, based in Sangin, 45 Commando was commanded by Lieutenant-Colonel Jim Morris, and held responsibility for a number of forward operating bases dotted along the eighty or so kilometres of the Upper Sangin Valley, as far north as Kajaki.

As with other brigade units, there had been no let-up for 45: on its return from *Herrick 5*, the Commando had completed a tactical simulation exercise and a winter deployment to Norway before returning to Afghanistan for *Herrick 9* as Battle Group North. Like the other battle groups in the area, the Commando's mission was to create the security conditions to enable the continued growth of emerging Afghan governance in the area. However, as 45 was split between five forward operating bases (FOBs), with each area presenting its own set of challenges, the range of circumstances faced by the CO and his men was unpredictable and often extreme. V Company, created solely for *Herrick 9* and disbanded on completion, was formed to protect the strategically important Kajaki dam. In the centre of the area of operations was the crucial Sangin district, secured by Y Company at FOB Inkerman, W Company at FOB Jackson and X Company at FOB Nolay. To the south, Z Company operated out of FOB Gibraltar, prohibiting any enemy movement between Gereshk and Sangin.

The Commando faced a determined insurgency but eventually suppressed the enemy's determined attempts to dominate and influence the lives of local nationals. These efforts often resulted in direct and lethal engagements but, overall, they were to create a stability that would allow schools, clinics and new shops to open: important indications of life returning to some normality for ordinary Afghans. They would also allow for progress in the training of the Afghan National Army and Afghan National Police, while always disrupting the narcotics trade in their area. 45 Commando also conducted a series of intelligence-led, targeted operations against the Taliban, each of which ended in the discovery of weapons, ammunition and explosives.

Battle Group North West, based in Musa Qal'eh and Now Zad, was the responsibility of the 2nd Battalion the Royal Gurkha Rifles, commanded by Lieutenant Colonel Chris Darby. Their secondary mission, for which they were held on permanent notice, was to secure the hold on the district centres in order to improve the Afghan Government's influence across Helmand, acting as a back up force wherever needed.

Battle Group Centre was the attached Danish Brigade based at Gereshk, while Battle Group South was the responsibility of the Queen's Dragoon Guards commanded by Lieutenant Colonel Alan

Richmond. Based at Garmsir, his tactical area of responsibility extended south to include Jugroom Fort, and north to Nawa.

The duty of 1 Rifles, 3 Commando's fourth manoeuvre unit, commanded by Lieutenant Colonel Joe Cavanagh, was in the OMLT role. This meant it mentored the 4th *Kandak* of the 3rd Brigade of the 205th Afghan National Army Corps. Cavanagh's battalion, a combination of the 1st Battalion Devonshire and Dorset Light Infantry and the 1st Battalion, Royal Gloucestershire, Berkshire and Wiltshire Light Infantry, was based at Camp Bastion, but with teams spread across the province.

42 Commando, commanded by Lieutenant Colonel Charlie Stickland, was detached from the Brigade's order of battle and based in Kandahar, under the command of Regional Command (South). However, the Commando's J Company (Juliet Company, but known as the Jesters and under the command of Major Alan Speedie), augmented by a platoon from 1 Rifles and a small team from the 2nd Battalion of the Princess of Wales's Royal Regiment (the theatre reserve, based in Cyprus), was assigned to force-protection tasks in Kabul and Lashkar Gah. This force protection in the west was to be undertaken from Forward Operating Base Argyll, where they were crammed into an old school house with elements of the ANA and an OMLT.

Argyll was to prove an early source of trouble for the incoming 3rd Commando: with no district governor to support, or to support them in turn, the thirty men of FOB Argyll were effectively in siege mode. It was a situation that was to shape the whole of *Herrick 9*.

As always, invaluable artillery and engineer support came from the Brigade's two permanent regiments. 29 Commando Regiment Royal Artillery, commanded by Lieutenant Colonel Neil Wilson, integrated joint fire and counter-fire within Helmand Province in order to protect coalition forces, target the enemy and help rein-force national governance. 24 Commando Regiment Royal

Engineers, commanded by Lieutenant Colonel Jim Weedon, influenced, supported and built, as needs demanded, by providing the full spectrum of engineering support to Task Force Helmand and the battle groups.

The Commando Logistic Regiment's Medical Squadron was largely reinforced by specialists from a wide variety of organisations such as the Territorial Army, field ambulances of all services and especially from individual reservists normally employed by the National Health Service. It was widely considered that there was – and is – more medical talent in the Camp Bastion hospital than in any large UK teaching hospital.

Among the smaller Brigade units were elements of 539 Assault Squadron, Royal Marines, who were deployed to Kabul to conduct ANA junior non-commissioned officer training, while thirty-six musicians of the Royal Marine Band Service covered a myriad tasks that centred round the field hospital, all the while providing much-needed entertainment.

In southern Afghanistan's Helmand Province, throughout the winter of 2006/2007, the men of Brigadier Thomas's Commando Brigade, conducting *Operation Herrick 5*, had taken on the Taliban face to face. The brigade had harried the Taliban across the featureless desolation that is known as the GAFA – the Great Afghan Fuck All – and had pursued them through the mud-walled compounds and tunnels of dense village complexes; among the rocky moonscapes and narrow defiles of the northern hills; between the dense bocage plantations and poppy fields of the green zone; and via the muddy, freezing melt-waters of the Helmand River.

By October 2008 it was felt that the time was right to finally ease up on the fighting and start concentrating on rebuilding and supporting basic government processes in Afghanistan: voter registration; poppy eradication; wheat seed distribution; and the building of schools and clinics, to name but a few. Easing up on the fighting was not to be so easily achieved as supporting basic government processes (although that was not so easy, either).

Voter registration and wheat seed distribution throughout *Herrick 9* were successful, although they did take considerable war fighting to achieve. During *Operation Herrick 9*, forty-two members of 3 Commando Brigade were killed, the highest casualty rate of all the *Herrick*s to that date.

CHAPTER TWO

LASHKAR GAH

One minute later, call sign Ugly 51 – the first Apache on the scene – reported another major body of Taliban on the river's west bank, about two and a half kilometres south of the Bolan Bridge. Suspecting them of heading north, accompanied by two Toyota Hiluxs full of weapons, the unmanned aerial vehicle was flown to investigate.

With the UAV keeping an eye on this new threat, the two Apaches flew down the river from where images of between seventy and eighty Taliban in arrowhead patrol formation were beamed back to the Brigade's terminals. Clearly well trained, these insurgents were also hand-railing, and visibly carrying RPGs and assorted small arms. This now became a properly indentified target and was immediately treated as such: after three passes by an Apache nearly all were lying dead, killed by flechettes and 30mm rockets.

Lashkar Gah – 'army barracks' in Persian – is the provincial capital of Helmand and the centre of the Afghan Development Zone. It lies to the east of the Helmand River and was established about 1,000 years ago as a riverside barracks for soldiers accompanying the Ghaznavid nobility to their winter capital of Bost. Bost and its outlying communities were sacked in successive centuries by the Ghorids, Genghis Khan and Timur Lang, but the ruins of the Ghaznavid mansions still stand above the river's banks. Nowadays, the area is better known for the new 7,500-foot aeroplane runway, the third longest in Afghanistan.

Across the Bolan Bridge, to the south-west of Lashkar Gah, lie the green-zone towns of Nad-e Ali and Marjah, with their outlying small villages, walled compounds and patchworks of productive fields. To the north is the Chah-e Anjir triangle and, to the south-east, Nawa. At its widest here, the green zone is nearly forty kilometres across, encompassing what was once regarded as the country's breadbasket but which is, currently, its opium basket. This fertility was encouraged by the many kilometres of irrigation ditches and canals built mostly by the Americans in the 1960s. The US programme was not just an economical one – they lined the provincial capital's streets with trees – but the programme was abandoned when the Communists seized power in 1978.

Following the comprehensive handover from 16 Air Assault Brigade, 3 Commando came to terms with their situation. As demanded by their commander, everyone across the brigade at every level would, in conjunction with the PRT's overriding aspirations, build on earlier successes. To do this, Brigadier Messenger required that the tactical initiative against the Taliban be retained through the application of a 'manoeuvrist' mindset; requiring his forces to keep on the move and to pop up in places unexpectedly, thus keeping the enemy guessing and under pressure. Coupled with that, he wanted his men's relationship with the Afghan National Army to be one of brothers-in-arms rather than master and pupil, a compelling distinction.

Addressing his primary concern, Messenger also demanded that all operations be focused on the local populace, in order to reassure them of their security by having confidence in the ISAF presence and thus in their own future. The counter-IED war would be an essential element of all his Brigade's activities and he would support anyone who exercised 'courageous inactivity' to save innocent lives, as his Royal Marines predecessor, Brigadier Jerry Thomas, who had first coined the phrase, had done. An example of this was when locals were first seen – through night-vision goggles – 'apparently' digging in the fields at a time of night when the observer could have been forgiven for thinking that IEDs were being laid. On further investigation, this proved not to be the case: the men were, in fact, the irrigation teams working to achieve the planned, nightly redistribution of water. It would have been a tragedy to have mistaken for insurgents, civilian workers who were simply going about their jobs. This was the sort of judgement call the friendly forces on the ground now had to make, but the feeling was that it was far better to miss the occasional opportunity to kill a member of the Taliban than put civilians at risk.

With prescience, Messenger and his team believed that the growing instability in the central zone – Nad-e Ali, Marjah, Nawa and Babaji – was the greatest threat to Helmand security. This had to be countered and resourced accordingly, but, significantly at this initial stage, it was an area not covered by a dedicated battle group. There were battle groups to the north and to the south but, apart from the small security detachment based in the beleaguered FOB Argyll at the very centre of Nad-e Ali, supplied by J Company of 42 Commando and an ANA detachment and its monitors from 1 Rifles, there was nothing.

Messenger's staff also identified highway security as another growing challenge to the civilian population, while safe conditions for voter registration had to be in place before January 2009. Security also had to be provided by the Brigade's presence for other projects, such as assisting the provincial governor with the safe distribution of wheat seed; the establishment of a Helmand-wide, civilian radio station; and the construction of the country's first purpose-built police station.

By the tour's end, many of these projects had been completed whilst others were handed on as 'works in progress'.

On the evening of 10 October 2008 at a few minutes before six, and two days into the transfer of authority, Brigadier Gordon Messenger was standing, talking, facing the combined Brigade and PRT staffs in Lashkar Gah's briefing room. Not yet in his stride, he stopped suddenly as a member of the intelligence team, trying not to run in the rush to get the news to those who needed it, anxiously pushed his way towards the Chief of Staff with a message. Clearly something was happening. The meeting stilled.

Groups of Taliban had been spotted coming from the direction of Nawa, to the south of the city, hand-railing (covering each other while manoeuvring in tactical formations) along the far river bank. That was unusual, and bad. Even more unusual was that further groups had been seen approaching Lashkar Gah simultaneously from the east, moving either side of a road towards an outlying checkpoint, manned by the Afghan National Police. Intelligence and the interception of enemy ICOM chatter suggested that their combined aim was a pincer movement on Governor Mangal's compound, before moving on to the jail and releasing as many Taliban prisoners as they could. In the process, they planned to kill as many of the ANA and ANP as possible, raise havoc in Lash and then disappear.

The Brigadier's briefing was suspended at once, while more reports swiftly followed. A second ISTAR message confirmed that between five and seven vehicles were moving north towards the river. Having come from the south, the ubiquitous Toyota Hilux pick-ups – the Taliban's preferred vehicle for ferrying men and weapons, often with a mortar base-plate in the back – were now driving along the western side of the river. Making no attempt at subterfuge, they clearly meant business. With this departure from the norm, it was serious business.

Authentication came through almost immediately:

1759:   J2 [the Intelligence Cell] confirm approximately ninety enemy have been positively identified carrying weapons.

Going by the stringent rules of engagement, this meant they were now legal targets: but who could take them on? They were in a brigade headquarters, not a battle group HQ, and thus had no one to coordinate Royal Artillery counter-fire, nor task helicopters: just a lot of staff officers with pistols! Nor were there as many boots on the ground as might have been, as the assigned protection force – half of J Company's Jesters – had been sent from Kabul to relieve B Coy of the Argyll and Sutherland Highlanders at FOB Argyll. There was, though, a large signal squadron at the HQ, and as marines are commandos first and specialists second, from that point of view the situation was containable. The Royal Artillery also accounted for twelve men in the main HQ, plus a few signallers, while there were also fifteen sappers who formed the counter-IED task force. All told, the Brigade HQ element was about 200 people, with about 60 per cent of them Royal Marines.

The reports, coming now in quick succession, carried increasingly disturbing news. The IX Group had identified a significant number of enemy armed with heavy machine guns and rocket-propelled grenade that were, according to intercepted ICOM chatter, going to attack the Bolan Bridge ANP checkpoint at 1915. Once they had neutralised that position, they planned to carry out a river crossing north of the bridge to attack the National Directorate of Security compound – the Afghan equivalent of MI5 – in Lashkar Gah. The time was now 1815.

The report ended:

This is consistent with previous reporting. It is credible that the ANP will be attacked.

With unconfirmed reports that Governor Mangal's compound, close to the Bolan Bridge, was taking incoming mortar fire, there was an air of incredulity among the Brigade and PRT staff, one that reinforced an initial sense of impotence and even of strategic shock.

To start planning, the Brigadier, his Chief of Staff and others moved to the Joint Operations Cell, from where an Apache attack helicopter had been requested as soon as the first hint of trouble had been received.

At 1853, nearly an hour after the first report, Neil Wilson, the gunner CO, was finally able to watch a download link from the Apache on his fire-control monitor. The apparent delay – in fact, it was a comparatively quick response from aircraft on Alert 15 – was the result of Task Force Helmand having no operational control over any air assets: all helicopters were 'owned' by Regional Command (South) at Kandahar, and thus required everyone to bid daily for their use.

As swiftly as they were able, two Apaches were on task, as was an unmanned aerial vehicle (UAV). Their combined reports indicated a fast-increasing threat, while other intelligence sources confirmed that the Taliban were intent on targeting the prison, the main bridge and the National Directorate of Security. A couple of suspected firing points were identified on the west bank of the river when three 107mm rockets were fired into the centre of Lash and two into Lash PRT; something that had never happened before. Thankfully, those that fell within Lash PRT failed to explode.

A couple of minutes later, a threat warning was issued by the J2 Cell:

1859:   Enemy forces are grouping in the outskirts of Lashkar Gah intending to target the headquarters of the ANA and ANP. Alternatively they will fight into the city. The attacks could start between 2030 and 2130. In addition, a significant number will look to conduct attacks elsewhere. All are armed with AK variants, heavy machine guns and rocket-propelled grenades.

One minute later, call sign Ugly 51 – the first Apache on the scene – reported another major body of Taliban on the river's west bank, about two and a half kilometres south of the Bolan Bridge. Suspecting them of heading north, accompanied by two Toyota Hiluxs full of weapons, the UAV was flown to investigate.

With the UAV keeping an eye on this new threat, the two Apaches flew down the river from where images of between seventy and eighty Taliban in arrowhead patrol formation were beamed back to the Brigade's terminals. Clearly well trained, these insurgents were also hand-railing, and visibly carrying RPGs and assorted small arms. This now became a properly indentified target and was immediately treated as such: after three passes by an Apache nearly all were lying dead, killed by flechettes and 30mm rockets.

The Commando Brigade had drawn first blood, but the overall situation remained serious and was getting worse as events continued to move quickly. A Taliban mortar base-plate position was spotted to the north of the Bolan Bridge, and several more large groups of enemy were identified west of the river. Then, at 1950, Governor Mangal confirmed that seventy further insurgents in Babaji and Basharan had ordered all local civilians to remain indoors while they moved towards Lashkar Gah. An Apache was diverted to verify this report but despite two passes over the area saw nothing.

Threats from the river's west bank were not the only ones, as reliable sources were speaking of between 1,500 and 2,000 armed men of fighting age planning to attack the PRT HQ and take the city. While this was a serious, almost overwhelming, threat, these numbers had to be taken with a hefty pinch of salt. As the Brigade's Intelligence Cell noted drily, the figures were the results of Afghan assessment: although the movement of fighters might have been credible, the numbers were likely to be doubtful.

What was not in doubt, however, was that an orchestrated, three-pronged attack was under way against the Governor, the prison and Lash PRT.

There remained another factor that was not in doubt either: it was vital that the enemy were engaged while they were still in comparatively open ground. If they were allowed to enter the urban areas, no weapon could be used against them due to risk to the local population.

All that evening, and well into Sunday's early hours, a succession of Apaches – relieving each other as they refuelled and re-armed – were kept busy observing, overflying in shows of force and, where certain of positive identification, engaging the enemy with Hellfire, flechettes and 30mm cannon. The enemy, however, were considerably dispersed across the countryside that night, and due to using the array of cover available to them, they were not easy targets, despite the pilots' thermal-imaging equipment and night-vision goggles. When uncertain of their target, the pilots would move away in order to tempt the Taliban to move out into the open again, but it was a time- and fuel-consuming, often frustrating tactic.

Fifteen enemy were killed close to the eastern ANP checkpoint, and another twenty on the east bank within a kilometre of the Bost runway. Far overhead, a B-1B bomber of the United States Air Force remained on call keeping up its own surveillance, but aircraft alone could not prevent this relentless build-up of enemy from three points of the compass.

Intercepting the Taliban chatter on their VHF radios, the ANP's bridge checkpoint reported that they were expecting a renewed attack between 0230 and 0300 hours. They were correct. An hour or so into the new day, loud explosions were heard from the direction of the Governor's compound. It was time to summon the

Brigade Commander back to the JOC from his quarters, where he arrived in time to hear that the bridge was now under sustained and substantial small-arms attack, coming simultaneously from the north-west and south-west sectors.

The Brigade log brought the evening's incidents up to date:

0122:   Enemy planning to target the Governor's compound with a direct attack.

Twelve minutes later another threat warning was issued:

0134:   The ANP control post west of Bolan Bridge is being attacked from the north by eighty enemy and from the south by fifty enemy. Approximately 150 enemy have travelled north from Kariz and are attacking the ANP in the vicinity of the prison. ANA commanders say that the enemy on the west bank have crossed the canal and moved to within 500 metres of the river in order to fire on the district centre.

Meanwhile, the worried Brigade staff could only sit tight within their compound, knowing that they were vulnerable to a full ground assault and certainly to attacks by mortars and rockets. This was not the start to their deployment that they had been expecting and, unable to combat either threat, the feeling of frustration was growing with each report. The battle was one between the Taliban and Afghan national security forces reinforcing and defending their static positions, with Lash PRT immovable in the middle, while the Apaches, circling overheard in the darkness, engaged every opportunity target they could.

0154:   Threat warning – enemy forces intend to move from the Bolan area to the Governor's house to 'take' and 'hold'. Afghan national security forces now moving south to Bolan Bridge to reinforce. Intelligence suggests enemy preparing to mount further attacks after removing casualties and redistributing weapons and ammunition.

In addition to the attrition wreaked upon them from the air, and although seemingly in great number, the Taliban were having to contend with quite a lot on the ground, too:

0157:   The ANA/ANP at the checkpoint to the west of the Bolan Bridge have repelled an enemy attack. While a significant number of enemy has retreated towards the woods one kilometre north of the checkpoint, there remains a significant number of the enemy one kilometre to the south of the checkpoint.

The Taliban had now established a position on the east bank, within the town's boundary and close to the Governor's compound, while a second team was preparing to use a culvert as cover; one that conveniently ran west–east, parallel with the bridge and half a kilometre to the north. This culvert cover would bring them to the river untouched by air and ANP fire, to a position from where they could link up with their compatriots on the town side. It was a bold move and one which if successfully pulled off, would double Taliban strength within the town. A UAV was tasked to confirm.

Meanwhile, seven kilometres to the east – the far side of Lashkar Gah – another Apache was observing 150 Taliban attacking an

ANP checkpoint that straddled the main road in from the east, from Kandahar. South of the road, yet more Taliban had abandoned their vehicles to hide among low bushes from the eyes in the sky. Unsuccessful this time, few survived what was to be the final aerial onslaught of the night.

At last, at 0405 in the morning of 12 October, the Taliban realised that their losses of men and equipment were outstripping their so-far non-existent gains. Instructions over their radios ordered the junior commanders to withdraw and head in a southerly direction. Those few survivors to the east of Lashkar Gah were sent back into the desert.

While the Brigade and PRT staff had largely been spectators to the battle, it was now their turn to reap the benefits of this incursion from the intelligence that was unwittingly exposed by the fleeing insurgents:

0416:   IX have eyes on vehicles leaving the area stopping and unloading bundles and people at various locations.

As they broke contact, the Taliban were unaware that a UAV was following them, all the while beaming back ghostly images to monitors in the JOC. Here, orchestrated by the IX Group, every compound and, more precisely, every individual building where arms were dumped or where men peeled off and faded into the night, was plotted with precise care on a large wall map. Each cache was marked with a numbered pin and the details noted down.

The Brigade couldn't engage from the air as these activities were among civilian buildings, but surgical strikes by commandos and soldiers in the coming days would be more effective and, as far as the dangers of collateral damage were concerned, they would also be extremely selective.

After seeing the product from the work the IX Group had done to follow the various groups all the way to Nad-e Ali, Litster was happy with the outcome of the night's work: the coalition forces had suffered no casualties and their knowledge of the area had increased immensely in just that one evening as they tracked the Taliban back to their bolt-holes.

It was twenty-four-carat intelligence. What was obvious was that the Taliban had a good system planned in advance for when they withdrew; a further sign of their being trained to a high standard.

Governor Mangal, too, was happy with the outcome, although he was less so when 1,000 Afghan national security forces personnel arrived, following a knee-jerk reaction in Kabul.

The Brigade had been under fire and the real colours of the local area had been displayed, in considerably more strength than had been expected. The old adage that no orders will survive first contact with the enemy proved all too true, and it was time for a serious tactical rethink.

During the night, two questions had been raised. Was the Brigade Headquarters in the correct place from which to conduct a complicated six-month, counter-insurgency campaign safely, untroubled by direct intimidation or, worse still, direct assault? Coupled with that, it was now instantly clear that the Nad-e Ali and Marjah problems needed sorting out immediately, but by whom?

Answers to the first question were considered, only for them to be dismissed one by one. The reason for this serious selection policy was that all grey areas needed to be addressed to pre-empt anyone from 'outside' seriously suggesting that moving the Brigade Headquarters was the correct option: nobody at Lash

Above: Helmand Province from the air.

Below: Fort near Dehe Su close to where suspected Taliban were thought to be hiding out.

Above: Kate Nesbitt (centre) helping to casevac a wounded Afghan soldier into a US Black Hawk.

Left: Chinook with an under-slung supporting Kilo Company during *Operation Sond Chara*.

Above: 45 Commando helicopter assault rehearsals for *Operation Diesel*.

Below: Chinook replenishment.

Left: Over-watch prior to the battle of Khan Neshin.

Below: Somme-like conditions on *Operation Sond Chara*.

Above: Chinook desert resupply during *Operation Aabi Toorah 2B*.

Below: Desert laager during *Operation Aabi Toorah 2B*.

Above: Route Somerset during
Operation Sond Chara.

Below: Lima Company, 42 Commando
patrolling in northern Kandahar.

Above: Headquarters trench Charlie Company. 1 Rifles controlling the battle after John List had been wounded.

Below: Poppy fields and *bocage* around Patrol Base Jaker.

Above: Stop and search during
Operation Sond Chara.

Below: *Operation Aabi Toorah 2B.*

PRT thought a move would be beneficial. In this view they were persuaded by history: in Iraq, during the autumn of 2007, the British had surrendered Basra Palace after it had come under regular, indirect fire, and retreated to the airfield, thereby, as some saw it, leaving Basra wide open. Following that decision, the British Army's reputation in America plummeted – something that could be understood by the officers now discussing the situation in Lash PRT.

They could see some pretty unhelpful similarities between the two situations: in Basra Palace, all the civilians had had to hide under their desks for minimal cover from mortar attacks. At the Brigade HQ, with no effective overhead cover, the occupants found themselves having to do the same thing. The Brigade also knew that if a civilian were to be wounded, they would all be made to leave by the Foreign and Commonwealth Office in London. If that happened then there would be no point in the HQ remaining there as, in effect, they would then only be providing security for the remaining military members of the PRT.

Already, the Brigade was talking about possible mission failure, in the same context that the British Army had failed in their decision by moving out of Basra Palace two years before. Added to this, everyone knew that the Americans were watching them carefully, fully expecting to see a second Brigade Headquarters being moved very quickly.

For the new military team in Lashkar Gah, to move away from the PRT was unacceptable. Instead, and with the lead-up to voter registration about to demand secure conditions across the whole of this heavily populated area, the only obvious, positive solution was to go on the offensive, into Marjah, Nad-e Ali, Nawa and Babaji.

To do that, and to answer the second question thrown up by the events of 11 October, the structure of Task Force Helmand needed to be reconsidered and increased by at least one battle group. This was the only option acceptable to Brigadier Gordon Messenger and his staff, and was one that was emphasised just seven days later by further indirect fire against Lash PRT.

The Brigade staff knew that their overall tasks, and those of the units for whom they were responsible, were to shape, clear, hold and build; yet they knew, too, that there was only so much territory in which they could achieve that with the assets assigned to them. Once the attacks of 11 October were over, they were going to have to enter some tough negotiations with the Permanent Joint Head Quarters back in Northwood to get the battle group they needed to cover the troublesome Nad-e Ali, Marjah, Nawa and Babaji area. The temptation was there for the Brigade HQ to tackle the problem themselves, but they knew it would be at the expense of all the other planning they had to do.

There was also, however, an escalating series of plans that could have been brought into play if things became seriously untenable. One was a move to an available, but limited in size, underground complex, while another was a move to Camp Bastion. However, that was seen as akin to mission failure again, as the staff would then be a helicopter ride away from the Provincial Governor; not to mention the message the move would be sending to the abandoned governor, and the Taliban, who would have succeeded in driving them out.

With the need for a new battle group, it was the Deputy Brigade Commander, Colonel Martin Smith, who stepped into the command gap. His unique background would prove useful in the intelligence war he would need to conduct in the search for and elimination of insurgents. After considerable deliberations within Task Force Helmand, he established what was loosely and temporarily termed Task Force Lashkar Gah on 2 November 2008, before finally establishing Battle Group Centre South.

Initially, the key players in the new battle group had to be found from among those immediately available. The first obvious candidate was 42 Commando's J Company, which was split between force protection tasks at Kabul and in the Lashkar Gah area: thus half this company was instantly ready. Within that company was also a platoon from 1 Rifles and members of the PWRR.

Before the events of 11 October, a major master plan to decentralise equipment and manpower had been well under way, with all spare equipment, personnel and intelligence being moved to the outlying battle group commanders. Battle Group North, for example – 45 Commando – had been given a few extra hundred men, artillery-locating radars, masses of ISTAR information and as much support staff as could be spared, to get on and deepen the areas that they held at that time.

However, on the fourth full day into the deployment they had to be reined back and told to get all the extra equipment back to the centre immediately: as a result of the decentralisation, there was now very little in Lash, yet it had just become vital ground and was to be protected at all costs.

The Brigade also needed to know, and quickly, what was happening in central Helmand. Colonel Smith and his brigadier knew that the new aim was to deepen rather than widen the Brigade's area of operations, although that too had to be modified to take into account the smaller numbers of men available to each battle group after the creation of the new one. The resulting message that was given to the COs was to pick one place within their area and concentrate on it as their vital ground for deepening: they could not be everywhere at once.

To find out what was happening in central Helmand was the duty of Lieutenant Colonel Andrew McInerney and his IX Group. He had realised that with the current intelligence resources in place and working, they could tell what was happening as far away as Sangin or Musa Qal'eh, but they didn't know what was going on just two kilometres outside the camp. Until that time, it hadn't been considered important, as resources had had to be siphoned off for the Kajaki dam project. As a result, it had become an intelligence black spot; one all too ruthlessly exploited by the Taliban just a few days earlier.

The IX Group was now able to help all the commanding officers to influence their individual areas by studying the human interest; the 'human terrain' as it was referred to. To help in this, they needed to get to know who the local elders were, and start building workable relationships with them. The local population were to be brought on side through *shura*s (meetings), influence and firm demonstrations of practical security. They had to be engaged.

Operation Herrick 9 had certainly started interestingly: it wasn't supposed to happen, of course, but the first unit into action had been the Brigade Headquarters itself. This firmly indicated that a re-think on previous assumptions and conclusions was needed, before the tour had even started.

The biggest positive outcome from 11 October and the formation of Battle Group Centre South was the conception of *Operation Sond Chara* ('red dagger'). An operation that was conceived almost by default, it was to be one that would set the battle scene for pretty much the next twelve months.

While the new Battle Group Centre South was being established, it was timely that Brigadier Messenger should issue his first operation order:

Op. Order No 1. 30 Oct. 2008

Mission: Task Force Helmand, in cooperation with the Provincial Reconstruction Team and Afghan national security forces, will conduct stabilisation operations in the focus areas in order to deepen the Government of the Islamic Republic of Afghanistan's influence across Helmand.

CHAPTER THREE

NAWA AND SHIN KALAY

The enemy to the south-east continued to be suppressed with bursts from the LMG and a Javelin – aimed at the middle heat source of the six enemy – plus well-chosen shots from the riflemen obeying the 'Watch and shoot' fire order; an order made much easier to follow at night due to thermal-imaging devices, excellent for 'pinging' enemy positions while they think they are concealed by darkness.

Then the order: 'Charlie fire team. Fix bayonets.' It was getting real. 'Right, let's go!'

Before the excitement in Lash PRT had even begun to die, a swift reappraisal was under way. The attack on the night of 11 October had not just been an operation mounted by the enemy against military targets of the Afghan national security forces; it had been a direct threat to provincial Afghan governance. When a repeat performance, albeit on a smaller scale, took place a week later on the night of 19 October, it was seen to be indicative of the contempt with which the local Taliban felt they could treat Governor Mangal and the security units in the area.

The Taliban had made a mistake, however. By declaring their hand so early and making the added mistake of doing so directly against Brigade Headquarters, they had committed a fundamental error that was to bring considerable retribution on to their heads: they had poked two 107mm rockets into the hornets' nest, and the hornets were about to stir.

With startling clarity the events of 11 and 19 October had thrown up unexpected voids in the Brigade's dispositions, plus the realisation that there were no reserves to fill those gaps.

There was, in particular, one void now known to harbour many Taliban. As 16 AAB's brigade commander had warned, there had always been a problem west of Lashkar Gah. Now it was confirmed; the towns of Nad-e Ali, Marjah, Nawa and Babaji were in a no-man's-land that existed along the boundaries between military responsibilities. This problem was the first watershed of the new tour, and needed addressing urgently.

The Lash PRT had been aware of the problem building over the months, but by and large had considered it to be narcotics-related rather than insurgency-related, and therefore one that tended to be seasonal and cyclical, and not permanent. This analysis of the situation had been reinforced by the knowledge that the Taliban had effectively bought the poppy franchise in the area: heroin has been produced in Afghanistan – as elsewhere in the Middle and Far East – over the millennia, and the country holds the position as the world's largest producer of opium. This was very much to the financial advantage of the re-emerging Taliban, and the subsistence farmers were paid by them to grow it.

Being adept at exploiting the friction and uncertainty created by the 'narco' problem, the Taliban had subverted that process into an insurgency centred on the Nad-e Ali district. When the first hint of a possible assault by 300 Taliban reached the PRT HQ on 11 October, some thought it might have been – as happened in Northern Ireland – a 'welcome back' gesture to the incoming Brigade but others, with more conviction, were certain the timings had been coincidental. Neither party, however, agreed with the PRT's estimation of the situation: this was not a straightforward problem of drugs barons feuding, with a beleaguered garrison stuck in the middle, as had too often been assumed. For many, it could be termed an insurrection.

The area was not a total military desert, however. When the Brigade took over Task Force Helmand, there was already a small group, between twelve and fifteen strong, of Argyll and Sutherland Highlands living in Nad-e Ali's old school house, just north of the market and the derelict fort. With them were their mentees of the Afghan National Army, whose warriors, although having been in battle, were artillery men at an early stage of infantry training and a low level of expertise. They in turn were supported by a small, fractured and ineffectual police force. Known as Forward Operating Base Argyll, the inhabitants had freedom of movement for less than one kilometre from the district centre: beyond that engagement by enemy fire was certain.

Hemmed in on all sides, Argyll was operating in the school buildings of what was once a thriving market town. Small, cramped, dusty and dominated by the partially crumbled walls of a nearby deserted rectangular fort; mud huts and hastily erected temporary buldings surrounded by the ubiquitous HESCO Bastion ramparts were home to upwards of 60 men – British and Afghan – at any one time. There were few creature comforts other than a bed space with a hanging, cloth wardrobe, and showers – when water was available – were taken from suspended jerrycans. The lavatories were in the open and food was from ration packs, with only the occasional luxury of fresh vegetables and meat for a few days each month following a delivery, which also brought mail. A generator blatted noisily for much of the time to charge the ever-hungry radio batteries and, just occasionly, powered a televison set or DVD player. Fine sand covered everything, and found its way into every nook, cranny and crevice – human and otherwise. It was not a sought-after posting.

Argyll was a thorn in the Commando Brigade's flesh from well before the transfer of authority. Brigadier Carleton-Smith of 16 Air Assault Brigade had always declared his interest in this area but, as he had explained, the Kajaki dam project forced on him by the ISAF staff had taken up too many of his assets for him to become involved.

It was a situation that the Deputy Brigade Commander, Colonel Martin Smith, and the Deputy Chief of Staff, Lieutenant Colonel Copinger-Symes – responsible for logistics – had started to analyse before they deployed. They had realised that there was not a lot happening in the Nadi-e Ali area in terms of infrastructure to be protected, and had started wondering why the half-company had been placed there – effectively cut off – and why the out-going CO was so concerned with the area.

Brigadier Carleton-Smith had become troubled after all the significant incidents in the area had been plotted, from which he deduced that nobody owned – in fact, nobody had ever owned – the ground, apart from the Taliban. The Afghan forces had folded on it, as it was dangerous and uncertain territory. Coupled with this, during the 11 October attacks, it had been made even more obvious that the majority of the insurgents were based in the area west of the Helmand River and, broadly speaking, in the Nad-e Ali area in particular, as the Taliban's 107mm rockets' firing points had been identified as being in Nad-e Ali.

3 Commando had also gained early experience of the area: when the Brigade's Chief of Staff, Lieutenant Colonel Al Litster, arrived in Camp Bastion in early October 2008, he found Colonel Andy Maynard, the Commando Logistic Regiment's commanding officer, pulling his hair out with frustration over the situation there, too. Maynard's task was to resupply Argyll, but he knew that this would be nigh-on impossible as the roads were too narrow – if a vehicle was hit it could not be turned round – while half his trucks could not negotiate the bridges. Resupply was difficult and dangerous, with neither aspect being completely alleviated when it began

to be conducted more often at night, with helicopters. Even with this help, the occasional convoy of smaller vehicles was still called for, and the journey in and out remained incredibly risky. The entire area was, simply, overrun with Taliban.

The persistent need for combat logistic patrols in this area, on top of the normal framework operations as laid out in the Brigade orders, was an issue that would keep Brigadier Gordon Messenger and his staff awake at night, and one that eventually decided the call for an overhaul of battle group formation.

Having been instructed by Gordon Messenger to form the new battle group, there were a number of actions Colonel Martin Smith had to take before the new formation came into being.

Unaware that remedial action was being planned, the Taliban, by launching their second attack on Lash PRT on 19 October while also engaging Argyll with small arms, had unwittingly sealed their fate. It was time for action based on a new set of aims: to reconfigure the task force; to conduct shaping operations across Central Helmand; and, having successfully achieved the removal of the Taliban, then consolidate for voter registration and poppy eradication.

In fact, Patrol Base Argyll was under fire daily, to the point where the Nad-e Ali company were almost bored of the interruptions – they were more of a nuisance than a serious threat:

13 October 2008

0940:   Contact. Wait out.

Small-arms fire and automatic fire had been reported being received into Argyll from two enemy firing points, one a kilometre away, the other rather less. Then the universal cry:

'STAND TO!'

'Not again, Sergeant Major!'

'Get your bloody kit on, lad ...'

All round the dusty compound with its al fresco 'furniture' made from the metal frames of ammunition boxes and the wire lining of broken-down HESCO Bastion containers, men scrambled across the dust, struggling into body armour and helmets as they made for their defensive fire positions. Some, woken from their first proper rest in days, grumbled and cursed: the thought of being hit was not uppermost in their minds; missing out on sack time certainly was. Those about to swallow their first hot boil-in-the-bag meal in some hours swore loudly and obscenely; one marine more loudly than the rest as the contents of his silver rations packet spilled across the sand. A running marine shouted a hurried, 'Sorry, mate!' but his hungry oppo could only stare at his bacon and beans now soaking into the dried earth and mutter, 'Fucking wanker!' before grabbing his own weapon.

0945:   Small-arms and automatic fire. Enemy positively identified at Grid XYZ. Observing.

1004:   SITREP Sporadic fire continues. ANP moving into ambush position of enemy force's location. Assess insufficient ISAF call signs to make aggressive move towards enemy firing position. Expecting situation to develop.

1056:   Enemy ICOM chatter received.

Now, they were discussing the location of friendly forces and their own possible extraction routes: 'They are moving against us. You should come back.'

1141:   SITREP ANP occupying ring of compounds to the
          east of firing positions. All quiet.

As the morning in Argyll dragged on towards the heat of midday,
life would return to what passed for normal in the beleaguered
outpost:

'You ruined my fucking scran!'

'Yeah, well, we had to chivvy – we were under fire! Here, mate,
have some of mine.' Two of the bulky, boil-in-the-bag packets are
tossed across. 'And give us your weapon, I'll clean it for you.'

'Cheers, oppo!'

Reconfiguring the Task Force would be impossible to achieve
overnight, yet it was necessary to show the Taliban that the
Commando Brigade was not prepared to be attacked with impunity.

Thanks to ISTAR and the large wall map sprouting coloured
pins, collated on the night of 11 October, it was clear where much
of the enemy's munitions were cached. While the prime aim was to
create the conditions for the removal of these Taliban forces west
of the River Helmand, no full-scale assault was yet possible, nor
ordered, other than a steady series of framework patrols to become
familiar with the area, and soften it up a bit.

On the morning of 20 October, J Company was ordered to
patrol out of FOB Argyll on a small operation nicknamed *Tor Soba*,
whose aim was to disrupt enemy rocket-firing points in the vicinity
of the walled compounds of Zarghun Kalay, four kilometres to the
north-east of the FOB. 'ZK', as it became known, was a village
with which the men of the yet-to-be-formed battle group would
become very familiar over the next two months: intermittent
patrols across the area were conducted as and when the company
commander felt the time was right.

*

Events began to speed up. On 2 November, when Martin Smith assumed responsibility for Task Force Lashkar Gah in order to establish Battle Group Centre South, J Company of 42 Commando and C Company Princess of Wales's Royal Regiment were immediately taken under command. All future attacks would now be met by the retaliatory planning of a battle group HQ with the wherewithal, albeit limited to begin with, to retaliate without begging for help from higher formations.

The new commanding officer then flew to Nad-e Ali for an initial recce, which further convinced him that FOB Argyll was to be the base for all future operations for the battle group. Two days later, however, the battle group was formally established at Bastion.

Siting the new battle group in Bastion had been forced upon Smith by the limited resources available. Although he felt that he would have the control he needed only if he was positioned in the centre of the proposed area of operations, he was obliged to set up his main headquarters at Camp Bastion because, at this stage, he had wholly inadequate communications and thus no control suitable for command in the field. Added to this, not only was he lacking the technology; he did not have a Signals troop to manage any when it did arrive: all he had was a colour sergeant signals specialist acting as the regimental Signals officer. Smith also lacked any logistics back-up; an overworked warrant officer filled the role of Regimental Quartermaster as best he could, but without a staff, his small team was having to leech off the logistics pool in Bastion.

The new manoeuvre unit began with a strength that was about 35 per cent Royal Marines, but Colonel Smith was even more pleased to count among his new command a considerable number of useful vehicles. He now had Scimitars, Vikings and the Estonians with their Finnish *Pavi* sporting .50 calibre mountings at his disposal. Together, these made an impressive total of fifty armoured vehicles.

Further realignment of the Task Force was necessary. C Company PWRR was now responsible for the Lashkar Gah area of operations, while roughly half of J Company remained under near-constant harassment in Patrol Base Argyll in Nad-e Ali, manning the operations room. There was no doubt in anyone's mind that this was the way to proceed: they had inherited Patrol Base Argyll, including the poor state of affairs there and, despite the daily battles and the risky resupply operations, they were managing to keep the base sustained. Combined with the attacks on Lash, it was recognised anew that refocusing efforts in the central area – the Afghan Development Zone – was the key to success in Helmand Province. Reinforcing Argyll – the district centre – was therefore vital, and was to prove the genesis for *Operation Sond Chara*, the operation that was strongly linked to efforts that enabled voter registration in the critical areas of the Afghan Development Zone, and the distribution of wheat seed.

Sond Chara was conceived with the clear aim of re-establishing government control of Nad-e Ali in order to set the necessary conditions for voter registration, which was due to begin on 20 January 2009. The plan was not to engage in military adventurism, but the removal – not necessarily the killing – of Taliban was an inevitable function of the operation. In fact, the aim had to be one of removal as Smith knew in advance that the operation would not and could not involve total clearance. It was a big area and he simply didn't have the troops to hold it.

It was a convoluted position for a commander to be in. Smith's main HQ was in Bastion. His forward HQ (operations) was in Argyll, in the Nad-e Ali district centre, and he had an even smaller, roving, tactical HQ that did not include, as it normally would have done, his operations officer. His Operations Officer – co-opted from the PRT – would have to remain in the main HQ in Bastion, for a better overview of operations while the Officer Commanding Viking Company acted as the Tactical Headquarters Operations Officer in the field.

Smith knew that the forthcoming operation would have to last for at least two weeks if it was to achieve its full aim. He knew, too, that whatever form the return of governance to Nad-e Ali took, it would need to be achieved through a deliberate, all-embracing, coordinated operation and not through a series of patrols, no matter how aggressive or inclusive those might be.

The operation order for *Sond Chara* declared that it was to be a '... *deliberate Task Force Helmand/Provincial Reconstruction Team operation to improve security and stability in the critical, population-dense areas between Lashkar Gah and Gereshk (the Afghan Development Zone). The focus of Phase 3C, Decisive Operations, is to clear enemy forces from Nad-e Ali, in order to set the conditions for Phase 3D, the immediate stabilisation of Nad-e Ali. This will focus on supporting and shoring up the Government of the Islamic Republic of Afghanistan governance through a functioning district governor, and community council, implementing the Afghan social outreach programme; managing voter registration; and improving freedom of movement for the poppy eradication forces.*'

No operation, let alone one on this scale and over such a protracted period, could be planned without a full appreciation of the ground. Nad-e Ali, at 2,600 feet above sea level, was a roughly half-moon-shaped area, its bulge projecting towards the west. At its widest, it was about thirteen kilometres from its eastern, straight-edge boundary that ran north–south for about twenty-five kilometres along the Helmand River and its associated canals. This curved western boundary ran from the north to the south-west before swinging back to the south-east, its limits marked by the Nahr-e Bughra canal until its very western extremity, where it split into the Trikh Zabur canal that then curved gently back towards the river. Where the canals divided, the Nahr-e Bughra continued south-west, marking Marjah's desert boundary. East from Marjah, and across about seventeen kilometres of open scrubland and desert, lay Nawa, another insurgent stronghold.

Nad-e Ali was an artificially rural area created by canals built by the United States in the 1950s and 1960s – Nad-e Ali was, in a way, similar to a huge gated community but one bounded by canals rather than walls. Thanks to the criss-crossing waterways that ranged from a metre wide to deep, flowing waters many metres across, vehicle movement and human movement was severely restricted. A rich area, it was – before the current fighting – probably the richest in the country. It was the country's breadbasket and an area known for its high rate of production and wide variety of produce.

Much of the country between the desert and the river was bocage of heavily cultivated fields, divided by long lines of trees and dense hedges. The land was flat and it was difficult to obtain a good all-round view except when lying on the top of a compound wall or building. Plans to send fire support groups off to flanking hills were out of the question as there was no high ground in the area: an unavoidable fact that was to have a devastating effect on tactics. One of the consequences of this was that the marines and soldiers were to spend a great deal of time fighting from the roofs – the highest vantage points they could achieve for better fields of view and wider arcs of fire – but, of course, there they were exposed and uncamouflaged. Battle Group Centre South lost four men this way as a result and it was testimony to the marines' bravery that they never thought twice about climbing on to a roof if there was no other way of putting fire down.

Demographically, the area was home to people from over thirty tribes. Because it was fertile and because of the US development, there had been much inward migration, with people always eager to explain that they lived happily together. Smith believed that if the Taliban and the current conflict were to be put aside, there would have been a good history of happy relationships in the area.

Within Nad-e Ali's half-moon, the ever-expanding population of about 30,000 was unequally divided into pockets of high density, and was centred on three walled *kalays* or villages: Khushhal Kalay in the west; Shin Kalay four kilometres to the north; and Zarghun Kalay to the north-east; plus numerous non-walled settlements such as Nad-e Ali itself, Luy Bagh, Baluchan, Chah-e Mirza and others. Being entrepreneurial, the inhabitants had, over the years, diverted the irrigation system further into the desert as far as where the land began to rise towards the western hills. From these outer locales, there were few crossing points into the *dasht* but those that were usable were, to a military mind, key points that needed controlling.

Having control of the crossing points into the desert was not a point lost on the Taliban: all canal or ditch crossings, whether into the desert or between fields, needed to be carefully checked before each could be considered safe to negotiate, whether on foot or by vehicle. Once the crossings were owned by the security forces, however, they would be able to begin searching everyone that used them for explosives and weapons, and possible intelligence material.

The state of the weather was also relevant to Smith's considerations. The official start of winter was 17 November, with the first rains due a week or so later, which would reach a peak in January. Snow on the plains was expected within this period, with the melt-waters running off the mountains well into March. All these predictions would be well founded.

*

While gathering together his troops, vehicles and supporting arms to mount one of the largest operations undertaken by 3 Commando Brigade since the invasion of Iraq, Smith and the embryo Battle Group Centre South began probing operations to gain essential intelligence. Concurrently, the Afghan national security forces were involved in detailed collaborative planning that involved from the very outset the Provincial Reconstruction Team, led by Colonel Haydn, who was permanantly on hand to advise on what effective stabilisation plans should come into effect the moment that *Sond Chara* was complete.

With selected areas within Nad-e Ali cleared of Taliban, Smith reckoned he could get away with a company inside the district centre to hold it, while one remained on the outer ring controlling the crossings into the *dasht*. Marjah was ultimately the target in the south, while Babaji was the northern target, yet there was only so much Smith could do in the time and with the men available: he had calculated that he and his men would need to clear and control 180 square kilometres of territory.

These initial plans included an early use of helicopters in a series of deliberate night-time, large-scale assaults. Earlier *Herrick*s had used helicopters, but mostly only for administrative moves; these after-dark missions at company-plus strength would be the most consistent, complicated and enduring use of aviation in the *Herrick* series to date and would establish a vital bench-mark for future aviation assaults by the RAF and army units. Using its long experience of such night-time, heli-borne operations, the Commando Brigade was able to show the RAF and army air squadrons how it was done. From that point onwards, the procedures were passed on to ground troops to the point that such operations swiftly became the norm.

*

Smith needed to prepare his proposed battlefield carefully, for he knew that if he went straight into the first major phase of *Sond Chara* by just conducting some type of clearance without shaping the area first, he would take a serious number of casualties and be forced to use more force – and high explosives – than he would have liked.

To execute this preliminary phase, Smith sent what he possessed of J Company, under the command of Major Reggie Turner, into Nad-e Ali on 18 November with orders for him to consolidate his base in Argyll. He was then to prepare the battle space in any way he saw fit, but especially psychologically: as many recces as possible were to be conducted in order to understand the people and their area, while establishing as much freedom of manoeuvre as was practical.

Smith's previous recce to FOB Argyll had enabled him to get a feel for the ANA men he had inherited there. Due to the enthusiasm with which the enemy was attacking the place, he had been able to get to know fully what he could expect from them, and use this knowledge to the advantage of his planning processes.

Originally designated to command the Lashkar Gah Operations Company, Major Turner had, the previous June, visited the provincial capital, following which he had been able to report back to his men, who were still on pre-deployment training, that the living conditions there were comfortable. There would even be good communications with home, so it would be worth them bringing personal laptops and DVDs.

While it is widely known that time spent in recce is seldom wasted unless the situation changes significantly, when it *does* change, it's usually a tough call. With J Company's relocation to FOB Argyll, instead of the relative comfort and safety of Camp Bastion, the lads would now be surrounded on three sides by enemy, with limited resupply and a guaranteed exchange of fire before dusk each evening.

On his arrival in Camp Bastion, Turner was briefed that one of his troops, augmented by a small detachment of the ANA, was to conduct a relief in place – a handover of responsibility – with B Company of the Argyll and Sutherland Highlanders in early October at FOB Argyll. This meant he had to gather all the necessary equipment and personnel, and move them out as soon as possible. The rest of Juliet Company remained split – for the time being – between their original task of supplying security at Kabul, from where some of the men had been recalled to the bright lights and sangars of Lash Vegas following the 11 October attacks. However, a month later, it was back to Argyll for them again, and unfinished business.

When a major unit has its role changed at the last moment, to one for which it has none of the 'mission essential' kit, it can be a nightmare but, thankfully, everyone swung into action, and the troop was able to move to Argyll in good order.

Within twenty-four hours of their arrival in the beleaguered base, J Company's 2 Troop and Fire Support Group had turned the old school house into a home that included the amenities of a junior non-commissioned officers' mess, a luxury unknown even in Camp Bastion. Inevitably with marines, even more important must-haves took form: a gymnasium with sand-filled ammo-box weights, and reinforced positions for the GMG, Javelin and GPMGs in their sustained fire roles.

Other significant events were also under way. Governor Mangal appointed Habibullah Khan as District Governor for the area; an excellent move as now Martin Smith was able to sit down and talk to someone who represented the local population. The area's first District Stabilisation Adviser (STABAD) was also appointed: Major

Jim Haggerty was a proactive, ex-army officer whose arrival brought an immediate meeting of minds. He was to get to know and work well with Habibullah, as he spoke a modicum of Pashtun; enough to guarantee an understanding of, and keep a check on, the formal interpreters.

His appointment, however, came with its own inbuilt drawback of the need for added security, as Smith knew that the worst thing he could do was to get his STABAD wounded or killed. As all movement out of compounds carried risk, it was only when the right protection was in place that Haggerty could go with Smith for extended periods to the district centre, while Habibullah spent time commuting between the district centre and Lashkar Gah, from his office in a dilapidated building within Argyll's ANP compound.

Apart from the military necessities of preventing further attacks on the provincial capital and the establishment of secure conditions prior to voter registration and wheat seed distribution, Nad-e Ali now possessed a workable quorum to guide its future. In addition to the district governor, the commander of Battle Group Centre South and his STABAD, there was also the commanding officer of the ANA *kandak*, Colonel Abdulhai and Colonel Abdul Satar, the Chief of Police. Every evening, they would gather to sit round the harsh light of a Tilley lamp, discussing what needed to be done and by whom.

It was essential that the marines quickly began to form an association with the local people. They also needed to strengthen their interaction with the Afghan national security forces. As the OMLT battalion, 1 Rifles was doing that well with the ANA, but there was more work to be done with the police. Coupled with this, Smith wanted to forge a relationship with the Taliban: a psychological one, as he was concerned that the ANA soldiers were more scared to leave their compounds than the Taliban were to leave theirs. J Company did exactly what he asked them to do. They expanded the patrol area, and made their presence known – and felt – in the area.

With elements of J Company redeployed back at Argyll, events for deterring and disrupting further indirect fire into Lashkar Gah could take off at a pace. As Smith's men initiated the first framework patrols and began to meet the locals, this was, at last, a task the marines could get their teeth into. With the coordinated use of ISTAR and supporting gunfire, the Jesters of J Company began pushing the Taliban back.

Yet they were not to have it all their own way. On the morning of 10 November, the Fire Support Group was patrolling towards the village of Zarghun Kalay with caution. They were still about two kilometres from its south-eastern corner when, quite unexpectedly, the group was engaged by the enemy at ranges of between five and thirty metres, with further enemy positions spread out for nearly half a kilometre. It was a well-planned and well-executed attack.

RPGs and small-arms fire that close are difficult to combat without the tightest possible command, and control of the group was being exercised by junior leaders.

'Shit me!'

'Take cover! Enemy right!'

'Seen! Do you want me to mark with tracer?'

'Yup, and make it hurt while you're at it.'

'Roger.'

Caught on the dirt track that lead north to Zarghun Kalay, Corporal Russ Coles realised that his WMIK was too close to be of any use in this fire fight. The enemy were close, with hand-held weapons easily brought to bear. His own armament, a .50 calibre heavy machine gun high on the vehicle, was almost impotent at that range and at the angle of depression needed to reach the nearby ditches. He had to get back, and put some distance between him and the enemy, before he could seriously engage them. However, as he began to move to create that necessary distance, he attracted unwanted attention:

'All call signs – am under fire. Watch out. I'm coming back.'

And then, moments later:

'Shit, the cab's been hit. Hold on, lads!'

Followed quickly by:

'Sixty-six!' – the cry from Coles for everyone to look out as he was about to fire an anti-tank missile, was followed again by: 'Sixty-six!' as a second missile was fired at near-point-blank range. Then an aggrieved voice:

'That'll teach Terry to mess up my wagon.'

The official log entry recorded back in Lash PRT was more measured:

1024: Contact. Wait out.

1043: Friendly forces contacted by enemy with small arms and RPGs. FF returned fire with GPMG and 2 x 66mm. Damage sustained to 1 x WMIK, extent of damage unknown at this time. No casualties sustained.

1105: UPDATE received. Call sign 10 now withdrawing to Grid XYZ. Call sign 20 providing fire support.

1112: Call sign 20 engaged heavily by five enemy, positively identified with RPG. Friendly forces returning fire with GPMG/SAF/66mm.

Once the 'troops in contact' signal had been made, an Apache was tasked to observe and produce positive identification:

1112:   Apache has been requested. Believe that EF intend
        to capture ISAF soldier.

1137:   Apache call sign Ugly 51 wheels up at Bastion.
        En route.

1154:   UPDATE Five enemy with RPG are moving north.

1202:   Apache call sign Ugly 51 looking into it.

1203:   SITREP received. Call sign 10 back at checkpoint.

1209:   All call signs complete at checkpoint. Troops in.
        Contact closed.

It had been a close call, during which time the chilling enemy
force's objective mentioned in the log report had been intercepted
from the Taliban's ICOM network:

'Praise be to Allah. Your ambush *must* capture an ISAF soldier.'
'*Inshallah.*'

Not only was no one captured but there were no casualties: a
remarkable example of cool heads on young shoulders, but also
a testament to their training, especially that of the corporals. Coles,
in the leading WMIK, had fired two 66mm anti-tank missiles to
cover the extraction, during which his vehicle was badly damaged
by enemy gunfire.

Added to this, as is so often the case, the incident brought to
light some good intelligence: the Company Intelligence SNCO
had noticed one telling departure from the norm, the presence of
children along the road. Their absence was usually considered to be
a good indicator of impending action, as locals would be warned to
stay away shortly before an engagement. With no regard for the
safety of the local population, this contact was a sign – correctly
interpreted – that suggested the fighters were from out of area.

*

Martin Smith was pleased with the way his opening moves were panning out. By now, the colonel had gained enough information, enough freedom of movement and created enough uncertainty in the minds of the enemy to consider his next, more kinetic actions. He had, too, the beginnings of an understanding of the local people, gained through a series of preliminary *shuras* – held in order to establish a mutual understanding and even sympathy – that discussed the minutiae of daily life rather than more detailed problems. Coupled with increasing goodwill from the police, the ANA relationships, delivered by the operational mentoring and liaising teams, were developing well, too.

Smith's meetings with the commanding officer and the second in command of the *kandak* – the COs were changing and so the second in command was running it – and the Chief of Police, Colonel Abdul Satar, were proving to be useful. Colonel Satar had been in the area for a long time and knew a lot, but was under the same pressures that all policemen in that area were: there had been a number of incidents with his police which suggested to Smith that there was an element of collusion with the Taliban. There were certainly drug problems within the ANP and, no doubt, some coercion and bribery too.

All in all, J Company had done a stunningly good job despite having very little time to prepare, and very few of the right resources, but their combined actions had managed to get the Brigade over the start-line of *Operation Sond Chara*.

The next phase and the start to the kinetic work that was necessary to achieve the overall aim of *Sond Chara* was *Operation Marlin*.

Despite 42 Commando being under command of Regional Command (South) and being based at Kandahar, events in Helmand had been demanding more and more men, and Lieutenant Colonel

Charlie Stickland had been called on for support, and help. He had already lost J Company to FOB Argyll and L Company to 45 Commando to support Op DIESEL, and thus only had his Commando Reconnaissance Force and the Black Knights of K Company to deliver the clinical raid of *Operation Marlin*.

ISTAR had identified possible narcotic factories and Taliban stores in a series of compounds, numbered for the purposes of the operation 1 to 7, between two canals running south towards Nawa. These, collectively, were given the name Objective Marlin. The aim was to surge in many troops very quickly, a shock tactic designed to make the enemy panic and divert their attention, thus preventing them from adopting a suitable defence.

Required to get to Camp Bastion from Kandahar for pre-operation briefings and training, the Commando Reconnaissance Force had to conduct a gruelling twelve-hour, 120-kilometre journey across the *dasht* and, more dangerously, along roads and over bridges. To avoid any traffic, they left at night, during the curfew, escorting a number of logistic vehicles with a force of eighteen Jackals, eleven Vikings, a lone WMIK, and their brethren from the ANA in their hardy Ford Ranger 4x4s. *Operation Sond Chara* was to see the CRF really putting the Jackals through their paces: throughout their 120-kilometre insertion trek, the previous hard-won lessons in mobility operations started to pay off and the 'bogged down in a large mud puddle' recovery process was to become a CRF speciality.

Meanwhile K Company and Commando HQ too began their journey from Kandahar to Bastion in the dark on the evening of 1 December. In contrast, they should have had a trouble-free flight in a C-130 Hercules. However, the RAF's insistence, unlike all the other air forces in the Afghanistan theatre – the Americans, the Canadians and the Dutch – on only flying into Camp Bastion in the dark had to be adhered to, and was to prove an infuriating aspect throughout 42 Commando's tour.

Whenever they were required to move to Camp Bastion, the lads of 42 would have to wait around in Kandahar, turning up at 0200 for a flight at 0500, only to land at 0530, thereby never getting proper sleep. It proved to be pretty grinding, and didn't help the traditionally suspicious relations between the army and the RAF.

This particular flight, however, was to surpass itself:

'Over here, lads.' A member of the RAF's Movements staff waved his arms. 'Put all your kit through the x-ray machine.'

'What the fuck for?'

'Any sharp instruments. Knives – that sort of thing.'

'What about my weapon and ammunition?'

'No. Just knives. Not allowed on the aircraft.'

'Bayonets?'

'Just knives.'

The puzzled marines summoned their company sergeant major. A few well-chosen and appropriate words from the CSM quickly informed the peace-time RAF team of the realities of a war zone, and all was eventually able to proceed smoothly. (In contrast to this, when another company was required to move between Kandahar and Bastion quickly, the immediacy of the move meant that an available Canadian plane was earmarked for which, by habit, the marines arrived the statutory three hours in advance. The Canadian Movements officer was delighted with their early arrival, and allowed them to leave there and then. By habit, the heavily laden and dangerously armed marines had begun heaving their kit on to the x-ray machines until the Canadian had remonstrated with them: 'What on earth do you think you are doing? Stop wasting my time!' Explanations had left him speechless.)

The commander of 42 Commando's K Company, Major Jules Wilson, held his Company Orders Group the next day at Camp Bastion, during which he stated that K Company's mission was to seize and exploit Objective Marlin in order to disrupt the enemy, assist in the dislocation of the insurgents across the central Helmand belt and degrade the enemy's long-term perception of security.

As with all military orders, the mission was short, stark and unambiguous, while it was the job of the intent and scheme of manoeuvre paragraphs to amplify the commander's overall aim: *This operation is a first-light raid to disrupt insurgent activity within the Nawa area of operations. My intent is framed within the belief that I will minimise risk to my force and increase the chance of mission success by concentrating effort on seizing target objectives as rapidly as possible. Once secured, I will exploit the enemy compounds and target area in detail whilst simultaneously ensuring I am balanced and prepared for any insurgent response, particularly ground reinforcement from the South and East plus indirect fire. I will then melt back into the* dasht *during darkness and re-constitute for subsequent operations as part of* Operation Sond Chara.

With the arrival of Major Adam Crawford and his Commando Reconnissance Force, dusty and thirsty, the two teams now spent forty-eight hours in battle preparation and procedures until 6 December, when an enforced rest was ordered, prior to embarking in four Chinooks at 0300 the next morning.

The pre-dawn helicopter assault on 7 December 2008, involving the Black Knights of K Company, in concert with the Commando Reconnaissance Force, opened the batting for *Operation Sond Chara*.

Chinooks, each lifting forty men from 4, 5 and 6 Troops plus the Fire Support Group and Company HQ for the twenty-minute flight to the objective – at least, that had been the plan. However, the serviceability of the aircraft had begun to cause concerns during the previous day's rehearsals, and from then until the time for departure, the number of available aircraft went from four, to three, to two, to four, then finally levelled out at three again. It was not an encouraging start.

Due to this problem, Jules Wilson was obliged to change his plan of attack. Instead of landing simultaneously at two landing sites either side of the objective, only the one, landing site Prague, to the north-east, would be used by 4 and 6 Troop, the Fire Support Group and the HQ company. 5 Troop, due to land at landing site Bratislava to the south-west, would now come in a second wave. With surprise compromised, their landing site would be offset in the desert to the west, from where they would move on foot to their objective. This was also close to the lying-up position chosen by Adam Crawford for his 185-strong Commando Reconnaissance Force, once they had completed another, more covert, approach across the desert.

The route Adam Crawford took led his reconnaissance force south from Camp Bastion directly into the desert, before starting a massive, westerly dog-leg to keep well clear of Marjah and any sharp-eyed insurgents there.

Once clear of the southern limits of the town, Crawford swung his convoy of 22 Jackals, a handful of WMIKs and Vikings – as ambulances and support vehicles – east then north into the middle of the Marjah–Nawa gap, to a position about seven kilometres south of the five-way junction between roads and canals known as Check Point 9, Banshee or Green 24 (the differing names reflected the changing nationalities of the ISAF troops that had patrolled the area since the start of operations in the country). The reconnaissance force's position was now equidistant – about seven kilometres each way – from Marjah and Nawa. Crawford and his marines now waited, hull down and hidden, at the beginning of what was to be a lengthy period in the desert. The Jackal-borne warriors of the Commando Reconnaissance Force were, in fact, to spend the entirety of December away from the homely comforts of Kandahar, operating from the desert and, among other things, utilising their Jackal's off-road ability to secure Commando helicopter landing locations, maintain operational surprise and act as a blocking force.

For K Company, the flight in a cramped aircraft carrying forty marines with all the paraphernalia necessary for a period of active service in the field was thankfully uneventful. Uneventful that is, if a twenty-minute ride sitting on an uncomfortable web seat or standing with legs and knees taking the full weight of armour and equipment while encased in a vibrating, windy, stinking, blacked-out airborne tube can be considered uneventful.

Unlike many vehicles, the Chinook makes it impossible to have any of the pre-battle banter that keeps morale high and thoughts on the immediate future properly focused. Immobile in its bowels, there is nothing to do but think, and thinking at such times can be counter-productive.

Sitting alongside Jules Wilson was his JTAC team, who were at least able to communicate – above the scream of the twin engines and distinctive *woppah-woppah* of the twin rotors' broad blades – with the accompanying attack helicopters. Through their night-vision goggles, the pilots of the Apaches were able to keep up a running commentary on what they could see on the ground as the small aerial fleet approached the helicopter landings sites of Bratislava and Prague: each one either side of Objective Marlin and about 500 metres away from it.

In accordance with long-standing practices and for the avoidance of any possible doubt on such finely tuned occasions, the troop commanders carried flash cards to pass to the loadmasters as they boarded. These gave their call sign, destination, helicopter landing site name, grid reference and, of the utmost importance – especially in the dark – an azimuth towards which the direction of the ramp should, by preference, be facing when they disembarked. If the wind dictated otherwise as they touched down, it was a simple question of the loadmaster shouting the amended direction into the troop commander's ear before he led his men down the stern ramp.

This time, however, all the aircraft landed precisely where planned, facing in the desired direction. Equally precisely, as rehearsed, at landing site Prague, 4 and 6 Troops deployed directly towards their objectives. Royal Engineers carried mousehole charges for firing holes in compound walls, while others carried short ladders. There were two ways into a compound: through a hole blown in the wall or via the roof.

Major Wilson had not planned for reserves as he had felt that an instant foothold, in strength, in the target compound, was more important than having lots of reserves hanging around: the priceless element of surprise would make up for any lack of follow-up troops. He knew, though, that *in extremis*, 5 Troop, who had landed in the desert at landing site Bratislava to approach on foot through the green zone and engage any enemy forces choosing to make a run for it, could be called in.

The Fire Support Group, who had also landed on the east flank, swiftly took up their allotted positions to watch through their night-vision goggles as entry points to the target compounds were blown with the Royal Engineers' mousehole charges and the lads scaled the first walls. Following Wilson's orders, the Fire Support

Group now began their surveillance over-watch to catch any 'leakers' trying to escape.

A systematic, thorough and highly productive search of the compounds continued until after dusk when, in compliance with his own instructions, Jules Wilson ordered a phased withdrawal via two routes, to link up with the Commando Reconnaisance Force. Before doing so, he quickly sent the statutory report to be filed in the HQ's log:

1003:   Significant find of weapons and IED components. 4 x RPG warhead plus launcher, 2 x AK 47, recoil-less rifle, 6 x HME containers (2 including Det Cord), 4 x 1ft IED pressure plates. Further items found in car.

1321:   SITREP Friendly forces currently holding 1 person of interest, who may become a detainee.

1540:   SITREP 90 x bags of opium found whilst conducting a search.

Once night had fallen they then slipped through the green zone into the *dasht* to rendezvous with the CRF. It had been a gruelling twenty-one hours. From there, they were recovered by air to Camp Bastion around midnight.

Meanwhile, the CRF remained in their desert lying-up position, waiting for the orders that would bring them into close and continuous contact with the Taliban at Patrol Base 9/Green 24/Banshee. Whatever name it went by, it was the five-way crossing of canals and roads that would mark the southern extremity of *Sond Chara*, and had to be in friendly forces' control at any cost.

As a simple, swift raid designed to dislocate and disrupt insurgent activity, *Marlin* had been a success. As a counter-narcotic operation it was certainly that, as the approximately 400 kilograms of wet opium found would have been worth about £2,000,000 on city streets, while a number of weapons and a small hoard of ammunition were discovered along with some interesting documents. Three suspected insurgents were detained and eventually passed up the line for questioning. Later indications suggested that these men had played major insurgent roles in central Helmand.

Thanks to *Marlin*, and daily shaping patrols, by 8 December all was ready for another of *Sond Chara*'s opening phases – the securing of the boundaries. However, as it was Eid the next day, in deference to local religious sensitivities, it was not an active one.

Colonel Martin Smith knew that the primary directions of concern were from the west and south, both areas bounded by substantial canals that, with their crossings into the desert, would need to be secured before any attacks on the villages could be entertained.

From Banshee, the Trikh Zabur canal runs north-west then north for twelve or so kilometres, to where it meets a junction with the Nahr-e Bughra canal that separates the whole of the Nad-e Ali/Marjah district from the desert. This northern, vital junction had been given the code-name Barbarian and was also a key point that needed securing. Patrol bases were to be established at both ends of the canal: 42 Commando's plan was for K Company was to take Barbarian in a night *coup de main*; L Company was to do the same on the canal, about halfway between the two ends; while the Commando Recce Force and Manoeuvre Support Group would creep forwards to take Banshee in a ground attack. The canal, with its associated levee running between the two major patrol bases, was named Route Somerset.

It is a fact that as soon as guns are moved around any battlefield, an intention is advertised. With B Company of the Estonian infantry and a battery of 29 Commando Regiment's 105 light howitzers now joining J Company, FOB Argyll was nearly full. To accommodate them, an extension had been added to the compound, made out of the inevitable HESCO Bastion walls.

The keenness of the Taliban's dickers was such that nothing much could be hidden from the enemy's network of eyes and ears, but it was still considered sensible to lift the guns in by air at night, with the immense amount of ammunition needed for them following by road (of everything needed for a battle, 75 per cent will always be ammunition). To address this problem further, Smith saw to it that as soon as the guns arrived, he had them supporting 42 Commando on a fire mission to the south, to give the impression that although ISAF had established a fire base in Nad-e Ali's district centre, their attention was focused on the south, and not further north, where his true intentions lay.

Colonel Martin Smith called his first orders group for the next phase of the operation. The manner in which Smith planned to take care of Nad-e Ali was governed by a few overriding thoughts. Unable to secure the whole area, Smith decided he needed to create bubbles of security based on the key centres of population. He wanted to do this with the least possible fuss, and with the least possible overspill from each target: the IED threat was a significant factor that influenced his planning, and although he needed to outflank each objective, it could not be by very much as it was vital to avoid lengthy approaches that increased the risk of hastily laid bombs. The taking of Shin Kalay first would secure a route into the desert, should that be needed, followed by capturing Zarghun Kalay simultaneously with Chah-e Anjir, two kilometres further to the north-east. With these two *kalay*s secure he would end up capturing Checkpoint Yellow Four, the most important canal crossing into the desert, just north of Chah-e Mirza. At this point,

Lieutenant Colonel Charlie Stickland would link up with Battle Group Centre South's HQ for a Battle handover.

The whole operation was designed to take about two weeks to complete as Smith knew that to take any longer would be out of the question: it was a Brigade operation, with resources very carefully apportioned. Nevertheless, he toyed with the thought of defeat and mused about what that would mean for the current campaign. In the same context, also, he wondered what he had to achieve to be able to claim a victory:

Did they capture and control a village? No. They wanted to hand it over to the Afghan government undestroyed and with its civilian population unharmed and safe.

Did they destroy the enemy? No. They were there to defeat the enemy.

Did they have a view on how many of the enemy they wanted to kill? No, that was immaterial.

Were they going to clear the area? No, they didn't have a third of the troops they needed to do that.

How could they leave behind as permanent the effect they wanted to achieve? There was risk inherent in the limited forces available to them and that continued security would rely on a successful partnership with the Afghan National Army and the Afghan National Police.

And finally: What would defeat mean? It would be reducing Taliban influence to the point where Habibullah became the pre-eminent leader within the district, able to govern on behalf of Mangal and the GIROA.

The best way to proceed in securing each village, it was agreed, was to gain control of the key locations, such as the mosques, markets and village 'squares'. Smith did not want the still largely untrained ANP to have too large a presence in the operations, as he wanted each action to be as sedate and as discreet as possible.

Thanks to Smith's previous postings within the Corps, the knowledge of what psychological uncertainty could do in the enemy's minds was well proven to him, and probably the most important aspect in this kind of warfare. He knew that after the objectives were taken and the bubbles of security were in place, the newly placed district governor, Habibullah, would be the key interlocutor with the villagers and therefore within Taliban minds. That was the way to beat the enemy: it was very much the psychological approach Smith wanted to generate and the thinking behind his planning.

Other forces also needed to be in blocking and flanking positions before *Sond Chara*'s main assaults could be executed. Well practised in the art, 42 Commando was chosen for this specialist task. Surprise would be achieved through two night aviation assaults to enable the building of the patrol bases and the security of Route Somerset – the western perimeter of *Sond Chara*.

At 1900 on 11 December, Major Jules Wilson and K Company, having emplaned at Camp Bastion thirty kilometres to the north-west, landed in the desert just short of their destination, Check Point Green 2 – shortly to be Patrol Base 1, also known as Barbarian. This was the vital crossing from the desert into the green zone. It was cold and dark.

The landing site, immediately 'inside' the canal, was positioned to one side of the objective on purpose to allow K Company to approach their targets under tight control. Immediately on touch-down, and before the dust and sand had settled – using it, in fact, as cover – 5 Troop pushed forwards to a compound that had been previously identified by ISTAR; one that was well situated to become their first patrol base.

For Second Lieutenant Tom Jenkins, the compound offered superb over-watch and he quickly handed the position over to the K Company Fire Support Group, commanded by Sergeant Ed Needham. The FSG included two sniper pairs, a Javelin Command Launch Unit with missiles and an impressive thermal sight capability as well as three GPMGs. Their efforts were instantly rewarded when a moped, trying to speed past the compound, was swiftly stopped by 1 Section. The driver was carrying a push-to-talk radio and had a hand grenade sewn into his waistband, and was soon identified as a well known Taliban dicker. Meanwhile 6 Troop, who had landed slightly further north, had practically landed on top of two further Taliban dickers, who, when quickly surrounded by Captain Gareth 'Fitzy' Fitzgeralds' men, abandoned their weapons and fled in to a nearby compound, pursued by Corporal 'Buster' Brown's section. There they were captured and detained. Round one to the marines and no casualties on either side.

Having exploited their prisoner for as much information as could be expected and then having sent him up the line for further investigation by the Afghan authorities (after which he would be released or charged), 5 Troop, with 2 Section leading, pushed south to its objective compound. Suddenly, beyond the canal cross-ing point – little more than a narrow foot bridge – three fighting-age males could be seen to the east through night-vision and thermal-imaging equipment. As quickly as possible, 2 Section went firm in an irrigation ditch while the rest of the troop took cover behind, with everyone observing through their sights. As no weapons could be seen the three men were allowed to continue, unaware of how close they had come to serious trouble.

Suspicion, though, still weighed heavily as, perhaps reacting to something due to the noise of helicopters, the three men now started to move more quickly. The marines still withheld their fire, but not for long. Another more belligerent-looking group was spotted moving across the marines' front, coinciding with an ICOM update.

'The bastards know where we are. They're trying to outflank us.'

'No way. Have you got positive ID?'

'Tellytubbies, all right. Small arms and RPGs.'

'Hoofing, mate. Zap the buggers before they go to ground.'

'Roger!'

A burst of small-arms fire and then:

'Contact! Wait out!'

'5 Troop in contact. 200 metres across the field to our front. 2 Section engaging.'

Then radio silence until:

'On you go!'

With 5 Troop leading, and 6 Troop in reserve, the advance continued for another 100 metres until 2 Section Commander, Corporal Tom 'Webby' Webster, pinged six men huddled together at the corner of a wall to the south-east. Scanning further to the south, he spotted four more armed men walking from left to right towards the wall. Silently, through well-rehearsed hand signals watched through individual night sights and goggles, Webby summoned Charlie fire team into cover just forty metres away from the enemy.

Stealthily, using all their fieldcraft skills, the commandos slid into position, thankful for their knee pads on the rough, stone-strewn hard earth. Still unseen, they raised themselves into fire positions. No verbal orders were issued nor needed: this was team-work at its most effective, with each man covering his oppo as they manoeuvred into the best positions, all the while conscious, too, that there might be other Taliban already off to their flank. Cat and mouse. Tom and Jerry. Think as the enemy thinks, but think more quickly.

Suddenly, the silence was shattered as the fire control order was given. Three riflemen and a light machine gun opened up without hesitation, an unstoppable weight of fire from out of the darkness. Tracer danced and ricocheted round the shocked enemy, but they summoned up the courage to return fire. Sensing good sport and identifying the marines' positions, the six Taliban to the south east joined in, peppering 5 Troop's positions with danger-close rounds.

These Taliban were well trained: the rest of the troop was now in contact from the original firing point manned by the six Taliban, and in less than thirty seconds the entire complement of 5 Troop was embroiled in a fierce fire fight.

The enemy to the south-east continued to be suppressed with bursts from the LMG and a Javelin – aimed at the middle heat source of the six enemy – plus well-chosen shots from the riflemen obeying the 'Watch and shoot' fire order; an order made much easier to follow at night due to thermal-imaging devices, excellent for 'ping-ing' enemy positions while they think they are concealed by darkness.

Then the order: 'Charlie fire team. Fix bayonets.' It was getting real.

'Right, let's go!'

With the rest of the troop ready to give covering fire, the team moved in line towards a wall to their front, where one fighter was found slumped forwards, very dead. A discarded RPG lay half-submerged in an irrigation ditch, its lethal warhead jutting obscenely from the muddy water. Satisfied they were not now in immediate danger, the team went firm while the dead man was searched; the RPG was left well alone in case it had been booby-trapped – the pin was already out. The dead man had been armed with an AK-47 and wore a chest harness holding six magazines. His weapon had not been fired.

As the rest of the section moved forwards, it was time to get on with clearing the enemy firing points to the south then seize 5 Troop's first objective compound. Charlie Team was swiftly re-rolled into Alpha and Bravo assault teams, supported by a gun team. Bravo was covered by the machine gun as they moved silently and, they hoped, unobserved, across the open ground. The enemies' heads were being kept well down by supporting fire as the marines crossed the darkened open ground to the compound, which they entered: thankfully, their new positions concealed them from the Taliban.

Once inside, the compound was systematically checked. The only inhabitants were the members of a scared family who were quickly reassured, albeit rather briefly due to the pressure of the circumstances. They had known the risks of staying, and now they needed no persuasion that they were in the wrong place. When it was all over, they would be contacted and compensated.

5 Troop held firm in the compound till around 0430, when it was time to move on to clear the surrounding area to keep the Taliban at a good distance. Progress was slow and methodical as every conceivable hiding place needed to be searched. A pile of hay bales outside one compound hid Lee Enfield rifles, RPG warheads, RPGs and pistols, all of which the bomb-disposal team destroyed. After twenty-four hours on the go, 5 Troop's patrols moved back to the newly constituted – but still unconstructed – Patrol Base 1.

4 and 6 Troops had not been idle either and during the following days, along with 5 Troop, the Company, under the ever watchful eyes and reassuring firepower of the FSG mounted numerous

patrols in all directions to clear compounds while establishing, as accurately as they could, the forward line of enemy troops. For some, this involved a freezing wait – while observing – in flooded irrigation ditches; Marine 'Aquaman' Hunt gaining the record after four before-dawn hours' immersion in freezing ditchwater up to his nipples.

As the sun came out, so did the Taliban. 6 Troop was contacted from a host of compounds dotted round a large expanse of open ground to the south. Their reserve, 5 Troop, pushed forwards among the nearest mud-walled compounds and the ones closest to the enemy firing points to the west. While one corporal adopted a non-manual rodeo-type fire position astride a wall with enemy fire zipping about him, the most useful task fell to another marine with his 51mm mortar and five high-explosive bombs: when the dust eventually settled, the enemy fire was noticeably more sporadic, but still needed to be silenced.

Second Lieutenant Jenkins called forward 2 Section: 'You know where the enemy are – 200 metres to your front. Across that open ground. Compound 13 is your target. Fix bayonets and go when you are ready.' It was a 'cold' task but 2 Section were fired up and ready.

Marine Jarvis of 2 Section, 5 Troop, was one of the commandos who broke cover with three others, running like crazy towards compound 13. With a 200-metre dash to the objective, he was twenty metres into his run when a hitherto unused enemy firing point to the south got busy.

'Shiiiiiiit!' he yelled as the ground in front and around him was peppered by a colossal weight of rounds. 3 Section was laying down massive return fire, but bullets – the dreaded Afghan bees – were still whipping round the frantically sprinting men.

'Right, sod this – with me, lads!' shouted Corporal 'Webby' Webster who, thinking as lightning fast as his feet were going, dived towards the nearest cover, from where he began to lay down suppressive fire with the rest of the panting section. None of them knew how they had arrived unscathed.

A few minutes of intense action later, and taking advantage of a lull in enemy fire, Jarvis and Webster made it the last few metres to the compound. They didn't know who was more surprised when they found a cowering civilian inside.

'Don't mind us!' grunted Webby. He got comfortable in the prone position next to the man, and calmly started to lay down an impressive number of rounds at the enemy.

Moments later, Colour Sergeant Watson came sprinting over to their compound, carrying his L96 rifle. For the sniper, the targets and ranges were easy, especially as he had a marine helping him spot his targets. Each time he aimed his shot through the Taliban's murderholes and one was seen to drop, the smile on his face proved he was one very happy sniper indeed.

Sporadic fire continued, with the marines finding it easy to ping the deadly murderholes: each time a shot was fired, a spray of dust erupted from them. Slowly, while continuing to fire from random positions, the Taliban began to retreat. With intense covering fire from 1 and 3 Sections, the remainder of the troop sprinted back across the open ground to safety. It had been a busy day and with last light approaching, it was time to regroup.

At the same time, and after three days lying hidden in the desert south of Banshee, Major Adam Crawford had moved the Commando Recce Force swiftly into the five-way junction, where a farmer was paid handsomely for his compound to be turned into Patrol Base 9 (or Stella).

The tricky bit for the CRF had been coming in from where they had been hiding in the desert to join the canal levee after last light, and drive along it as quietly as possible with thirty-three vehicles, including Vikings and Jackals. They also had to stop and dismount often to check for IEDs. It was Sergeant Adrian Foster's job to creep ahead in the dark to check for roadside bombs and other obstacles, his sixth sense on high alert at all times.

Eventually the unit reached its destination at the road and canal junction, unchallenged and undetected. Being a prosperous area, the compound they had commandeered was large, with a walled garden alongside fields of crops. A mosque stood nearby.

The Taliban were quick to warn the CRF that this was most definitely their own territory and that their presence was very unwelcome. To make their point more strongly, the enemy broke with their usual routine and began to initiate contact in the dark. They began patrolling in the open, close to the marines' positions until, at 0246 of the morning of 12 December, they engaged with small arms and RPGs from two positions, 200 and 800 metres to the west. The marines, who were still reinforcing their firing positions by filling sandbags and establishing their 360-degree defences, returned fire rapidly towards a small walled area, behind which the enemy was taking cover.

Beneath a constant stream of tracer in both directions, and with the incoming rounds thudding into the hastily prepared sangars, Adam Crawford and the company sergeant major, Warrant Officer Adrian 'Spider' Webb, kept themselves busy checking that all was well with the outlying positions. Their casual, morale-boosting pose was rather shattered, however, when with the shout of, 'RPG incomer! Take cover!' the pair landed swiftly in the shelter of the nearest ditch, lovingly entwined in each other's arms.

It was, though, not a time for laughter but air support. Within twenty minutes, two Apache attack helicopters – call signs Ugly 50 and Ugly 51 – were overhead, engaging with their 30mm cannon, their timely arrival complementing the three Javelins fired at enemy positions. However, with no likelihood of a resupply until a convoy from Bastion had made its tortuous way down Route Somerset's levee, and with any local helicopter landing site too 'hot' for use, the firing of the effective Javelin missiles had to be carefully controlled and accurate. But the team's snipers also took their toll, the successful head shots marked by the tell-tale bubbles of red spray as a skull disintegrated; through night-vision goggles it showed as a mottled green colour. By 0346, it was judged to be all over for the night, and the helicopters were released back to Camp Bastion. The battle for the Nad-e Ali district had begun in earnest.

The sniper team was not prepared to sit as targets themselves, so over the next few days their commander decided to take their brand of warfare direct to the enemy. He set up several sniper ambushes, inserting and extracting his men, by night, into compounds behind the enemy firing points, then engaging the enemy at first light.

The enemy snipers had established themselves early on top of crumbling roofs, and the true meaning of accurate fire was now discovered as the majority of enemy firing points were found to be murderholes in compounds up to one kilometre away. These were suppressed, while others in the team notched up an impressive number of confirmeds with help from their number two. Two team members also returned to basics by constructing a hide in a straw pile in one compound, while a further two successfully neutralised one insurgent from 950 metres. The hard-learned principles of camouflage and concealment paid off as the enemy was unable to locate their ambush positions, despite dispatching dickers to search for them, and victualling up random compounds in the area.

Meanwhile, L Company, with 3 Troop of 24 Commando Engineer Regiment providing close engineer support through route clearance, explosive entry into compounds and experienced 'rummage search' teams, had assaulted on to the canal then moved in convoy along Route Somerset to set the conditions for another secure base to act as a block to cut off all the routes from the south, and from where a troop of 29 Commando Regiment's 105mm light guns could be brought to bear on opportunity targets. It took eighteen hours of nerve-wracking, non-stop work to achieve the patrol bases's chosen position before the construction could begin, during which time it poured with rain, turning their efforts into a vast quagmire. Everyone's sense of humour was more than tested.

With 42 Commando safely bedded down in the southern sector of Nad-e Ali in their new patrol bases, and reports that insurgents were definitely disturbed, all was in place for the securing of Nad-e Ali's *kalay*s. The initial battle had taken the enemy completely by surprise thanks to the swift audacity of the commandos, coupled with the long-range, high-precision fire from the Danish Leopard tanks. In a final and desperate attempt to counter-attack, the enemy had fired 107 mm rockets at the tanks and reconnaissance vehicles, only to be met by withering return fire from the tanks, mortars and Javelin missile teams: a total of thirty-one 120mm Rheinmetall L55 tank rounds alone had been fired into the enemy positions.

The first phase of *Sond Chara* had been a complete success. The Commando was balanced and ready should anything unexpected occur.

The Taliban now went quiet. Maybe they could read the weather better than the marines, as the relentless downpour already affecting the construction and reinforcement of the patrol bases was to last for very many days, far out-pouring anything anyone had ever experienced, even on Dartmoor. Patrolling in thigh-deep mud to ensure the ground around the patrol bases was kept clear of enemy was hard, but the lads simply got on with the job.

Getting on with the job also included a move down Route Somerset to the southern patrol base – Patrol Base 9, or Stella – which had been established by the Manoeuvre Support Group and Commando Reconnaisance Force. From there, the CRF supported K Company's move back to Camp Bastion for a day of admin and scran, which would put the reconnaissance force well in place to then escort a logistic echelon back across the desert to Patrol Base 9.

Making the patrol base a permanent one took masses of defence stores, but once 59 Regiment Royal Engineers had finished their construction, almost always under sporadic fire, the CRF then extracted back into the desert for their next tasking.

On the first full day after the night of 11 December, while waiting for the logistic echelon to arrive so that Stella could be turned into a HESCO fortress, contact was continually maintained with the enemy. The convoy had been due to arrive under cover of darkness but a Viking and, awkwardly, a recovery vehicle, had both rolled off the levee into the canal, which was about five metres wide and ten feet deep.

The knock-on effect of this difficult recovery was a two-day delay in the fortifications of Stella, during which time it was impossible to push the Taliban out any further. There was, effectively, a stand-off between the Manoeuvre Support Group with its snipers, and the enemy, whose increasingly accurate fire was sharpening wits and deepening sangars. It was, however, thankfully the sandbags that suffered the most from this deadly attack and not the men, other than an engineer who was wounded in the ankle. The marines, however, were not amused to be welcomed to Stella by a volley of RPGs, swiftly followed by tracer, at a time when 'Terry' Taliban should have been tucked up in bed.

With no proper base to move into until Stella was built by the sappers, the shellscrapes became deeper and deeper. But morale was always high, especially following a five-round, 'Fire for effect!' fire sequence that, each time, silenced the opposition, while the sergeants' orders for another layer of sandbags was never ignored. The marines were even happier when, with the eventual return of Recce troop and the Vikings escorting the logistics convoy, the Commando Reconnaissance Force was complete again and the beleaguered base was re-stocked with Javelins and grenades. Even better, after a few days, they were finally able to move inside the newly built base and into their freshly-dug 'pits'.

When the building of Stella started, and as it continued, it was possible to patrol into the surrounding area. The third day of Stella's existence was by far the busiest, with relentless incoming fire. Each time a firing point was identified the fire fight was won, but the Taliban were a canny foe and it could take some time for the firing points to be spotted and neutralised. By the end of the week, though, every enemy position had been identified, cleared and destroyed by bar mines laid by the remarkably brave Assault Engineers and a handful of equally courageous sappers.

'Lightning' L Company from 42 Commando operated in the area between the two fledgling patrol bases, clearing insurgent routes and, with their partnered Afghan security forces, spread a coalition presence across this previously hostile terrain. All the while the Commando had a mechanised camel train ferrying stores across the desert to reinforce and sustain the bases. The Royal Marines and the Afghan National Army then carried out joint patrols to dominate the ground and to make contact with the local population in order to reassure them, and make clear their intention to bring enduring security and stability to the area.

This next step – carried out by contacting the local elders and talking to them – explained 3 Commando's presence in Helmand,

and what they were seeking to do in the whole of the Nad-e Ali area over the coming months. From an ISTAR point of view, voter registration and wheat seed distribution would be on the main agenda for this first *shura*.

Everyone was extremely pleased with the progress made, and especially so as it had been at no cost in 'friendly' blood. This success was put down to supreme field craft, boldness and outstanding fitness, which must always count. There had too, perhaps, been an element of luck – there always is.

Back at Battle Group HQ, Martin Smith knew that his blocking patrol bases were now in position to attack Shin Kalay. In order to get the supporting vehicles through, his first priority was to capture a tactically important bridge that lay 400 metres short of the village on the west-running Pharmacy Road. Taking the bridge intact was vital, as without it there would be a very long detour through difficult, close country that was perfect ground for quickly laid enemy ambushes.

Once it was in his hands, Smith's plan was for C Company from 2 PWRR and the Recce Platoon from 1 PWRR to hook on to the south-west corner of Shin Kalay and establish themselves 500 metres from it. Here they would be able to see anything or anyone coming out of the back end of the *kalay* – the so-called leakers. Then, with those arcs covered, J Company would secure, rather than capture, the centre of the village.

Shin Kalay had to be the first target. Not only was it a mere two kilometres to the west of FOB Argyll, but each evening the enemy was occupying a line of compounds between the village and the district centre, from where they would open fire on the FOB. The fire would always be returned and, after a brief exchange, quiet, of a sort, would once more descend, but it was an irritation and denied J Company full freedom of movement. It was necessary to push the enemy back.

A few days earlier, a small J Company patrol had reconnoitred as far forwards as Pharmacy Bridge. Known as Blue 25, the bridge crossed a substantial canal with, on the far side, an abandoned but intact pharmacy (it was in this whitewashed, pockmarked-by-rounds building that five British soldiers were to be murdered and six wounded in November 2009 by a rogue Afghan policeman, who had been planted within the service by the Taliban). Further west, across the flat, cultivated fields divided by wispy, dry hedgerows and irrigation ditches, brooded the beige-coloured mud walls of Shin Kalay. This final approach to the objective would be regarded as 400 metres of killing ground.

Martin Smith weighed up his proposed actions. His men had been down to the pharmacy, getting used to operating in the dark over that ground. The Taliban had carried out one or two concerted night attacks in the general area but usually, by midnight, they would draw back to their compounds: some had families to go to.

As Blue 25 was the key to any approach towards Shin Kalay, J Company despatched a second, confirmatory recce patrol at 0300 on the morning of 11 December. The members returned safely. There were no enemy in permanent residence.

D Day for the start of *Sond Chara* was 12 December with H Hour – the taking of Blue 25 and the pharmacy by J Company – scheduled for 0600. Shin Kalay itself would be secured on orders, subject to the situation, later in the day.

Prior to this, Smith had streamed a further two companies into the district centre but had immediately dispersed them to avoid telegraphing the presence in strength to the enemy's dickers.

At 0100 in the morning of 12 December, Colonel Martin Smith gave his confirmatory orders for the first phase of the operation.

Well before first light, a classic fighting patrol led by a Royal Marines colour sergeant lay hidden to the east of Blue 24.

With no sign of any enemy, the bridge and the pharmacy were in their hands without a shot having been fired by 0600. So far so good: but where were the enemy? Official and unofficial estimates gave figures of about 200 Taliban fighters in the area, but it was known that they were constantly on the move in an attempt to give themselves a bigger profile and make themselves look more numerous than they were. There were believed to be a further 100 fighters in Zarghun Kalay to the north, but the truth was that they were probably the same 100 flitting between the two *kalay*s on foot, mopeds and the occasional 4x4 Hilux pick-up truck, as befitted their purposes.

The group in Zarghun Kalay was regarded as a pretty professional bunch, coming south to put pressure on the north edge of the area. These irritations had involved J Company in regular, daily fire fights, and Martin Smith had wanted to teach them a lesson. This had led to some 'hoofing' actions.

The lads of J Company would lay ambushes before moving a patrol forwards to be contacted by the enemy. They would then withdraw, expecting the enemy to follow them and be drawn into the ambush. For some reason, the enemy hadn't always done that – they were wary of the WMIKs, and perhaps were learning lessons. In these instances, the marines would then move back into the killing area to contact the enemy even more closely than before, to then drag them into a fire fight while moving back to their own killing area. At that point the ambush would be sprung and the enemy killed.

Characterising the Taliban is always difficult. Many of those the marines faced were merely disaffected youths who would be given a weapon, told to go and fire at the enemy then come back when the magazine was empty. It was the hard-core foreigners who had to be watched.

J Company had managed to secure and dominate the Pharmacy Bridge area on the morning of 12 December without incident, but it was to prove, factually and metaphorically, a false dawn.

At 0658, only an hour after the bridge and pharmacy had been safely taken, an Apache attack helicopter received a troops in contact warning. Almost immediately, the pilot reported seeing RPGs and tracer being fired into the pharmacy from one of many fortified compounds along the eastern edge of Shin Kalay's perimeter wall. The enemy was subsequently engaged by the Apache's 30mm cannon, the first exchange of fire in a contact that would last for the next 100 minutes.

Delighted that the vital crossing had been captured not only intact but without casualties, Smith was not surprised by the enemy's reaction. While C Company of PWRR in their Vikings, after successfully negotiating a fire fight initiated from the south of the village, moved round to the south-east to suck up any leakers as planned, Smith decided it was time he joined his forward troops from where he could better control the battle.

Smith quickly established the Tac HQ in the pharmacy where, although he didn't know it at the time, they were to face an hour and forty minutes of fighting. With fire from the village passing over their heads, and unable to see their targets from ground level, a few of the Jesters decided to climb on to the exposed roof of the pharmacy; a dangerous position but an effective one.

Keen to be in full control, Smith joined J Company's commanding officer, Reggie Turner, on the roof. The battle group commander knew he had to win this fire fight before he could even think of sending J Company forwards to Shin Kalay. Overhead, US Cobra and UK Apache helicopters took it in turns to offer air support, and there was fixed-wing support from American F-16s.

Support with Hellfire and cannon was not the only thing the Apaches could offer; they were, in particular, tasked to observe. Smith was careful in his use of air cover and high explosives, and tended to use the 'eyes in the skies' for ISTAR, rather than the shows of force needed earlier in the Afghanistan campaign. Both the naval and the air force pilots became happy to conduct ISTAR operations, rather than concentrating on using the arsenal of ordnance at their disposal: they had taken on board the fact that they were in a counter-insurgency battle, so were less keen to 'move mud' and more keen to start to contribute in other ways.

As if to prove that this job was almost as vital as their kinetic support, one sharp-eyed Army Air Corps pilot suddenly reported a large number of women and children entering a nearby building. Thankfully, the pilot had been looking in that direction at just the right moment. This was always crucially important intelligence as the harming of just one child, while always appalling, could also put the battle for hearts and minds back by months.

Chillingly, the Taliban knew that too, and often used the knowledge cold-bloodedly to their advantage. Allied to this was their often uncaring attitude towards their own casualty figures, re-manning seemingly effortlessly from a huge pool of volunteers from a vast global collection of Muslim extremists while, in the case of local nationals, from many who were less interested in idealism than in remuneration.

Nor were the Taliban picky about using young teenagers to swell their numbers, as two casualties were to prove:

1343:	Two local nationals (males aged eleven and thirteen) brought in with gunshot wounds to abdomen and leg. MERT authorised at 1408 to pick up from HLS.
1421:	Wheels up from Bastion.
1441:	Wheels down at Nad-e Ali.

Just three minutes on the deck was all it took for the well-practised handover and acceptance of casualties to take place:

1444: Wheels up Nad-e Ali.

A twelve-minute, low and fast flight until:

1456:   Wheels down at Bastion.

After 100 minutes of trying to win the fire fight against a tenacious and determined enemy, the rate of incoming ordnance at last began to die enough for Smith to send forward his Jesters. They attracted yet more fire but as it was now more and more desultory – even with J Company trailing their coats – and they were, quite unexpectedly, able to move into the centre of the village.

Having secured Shin Kalay, Colonel Smith called the local area governor, Habibullah, with an invitation to join him in the village square. Within two hours of the fire fight, the first of the village's *shura*s was being held between the govenor and the village elders who, with their families, had been hiding in tunnels or had fled beyond the village walls.

While the elders and Habibullah's small team sat, cross-legged, along a wall, and the women returned to their homes to prepare *chai*, Colonel Smith's medical and dental teams began their very specific approach to the hearts and minds issue. Old men and women came forward with grandchildren. Stethoscopes were brandished, the young encouraged to 'listen in'. Temperatures were taken, eyes and ears studied and the odd tooth extracted. Sweets and chocolate were distributed. Slowly curiosity overcame the earlier, natural, suspicion and fear. Smiles reappeared, even laughter.

As civilians had been caught up in the fire fight (despite as many precautions taken as had been feasible, without sacrificing surprise), the inevitable questions would need to be asked, but on the whole, everyone was extremely pleased with the outcome. The hope had been to hold the *shura* in the mid-afternoon before it got dark, and this had been achieved. It was a great feeling for everyone concerned.

Once safely in Shin Kalay, and having held the first of a number of successful *shura*s, Smith; his stabilisation adviser, Jim Haggerty; and Habibullah decided to put in place a system that involved, each time a village was secured, at least two *shura*s.

The first one was termed the security *shura* at which Habibullah would lead, explaining what ISAF troops had done for the people, why they were continuing to do it and thus why they, the village inhabitants, had to help with keeping the Taliban out. Habibullah was always careful to explain how the locals were an essential part of the security apparatus and that their security could be guaranteed only if they were willing to make it work. It was, in fact, a chicken and egg problem: in a counter-insurgency situation, a protecting force can provide full security only if people are feeding them intelligence, but people tend to provide security only once they are feeling secure enough to do so. By always making the point for the need for mutual support clear, Smith hoped to be able to begin a new relationship in each *kalay*.

The second *shura* was often the more interesting of the two as it dealt with more local, social matters. It included any follow-up medical treatment needed, and often long discussions on seemingly unimportant matters; but all subjects that helped increase trust, and lessen scepticism. Smith and his men knew that no subject raised by the elders was too trivial.

The villagers' response tended to be guarded, as fear and suspicion remained close to the surface. There were still Taliban in the area; there would always be Taliban in the area. There were even always Taliban at the *shura*s. Smith knew, for instance, that two of the local 'elders' with whom he enjoyed a cup of tea in Shin Kalay were local Taliban, interested in why he was there and what he was doing. He knew, too, that for an elder to declare in public and too readily his acceptance of, and gratitude for, the presence of ISAF forces was a foolhardy error that would, without any doubt whatsoever, lead to reprisals from the Taliban, both individually and collectively. It was not an enviable position to be in.

The plan for the first *shura* would from that day, Smith hoped, follow a set routine. Habibullah would speak, and would then hand over to Smith, who would then pass the lead to the CO of the *kandak*, who in turn would hand over to the Chief of Police, with everyone having their ten minutes' worth.

Smith, in particular, always had to speak carefully at these *shura*s. He knew he needed to be totally honest with the local people, at the same time as there were certain things he needed to impress on them. He did not want them to think, for example, that he was in their village as an imperialist invader, and would insist on making it quite clear that he wanted no part of Afghanistan. His family, he would explain, were back in UK; he was in Helmand merely to do a job and when that was done he would then return, with his men, to the United Kingdom. Everything, he insisted, was done with the top support of Habibullah and, through him, the Provincial Governor in Lashkar Gah.

Smith was helped in this task by an ace up his sleeve. He had needed a way of making sure that he had the right cultural advice, and had made sure he had direct personal connections with Afghans who could speak English. By luck, also, he found a Territorial Army lieutenant called Mike Martin, who was running the interpreters in Lash and who had completed a fifteen-month Pashtun course. He was very well attuned to the culture and became Smith's adviser.

Martin was to be essential in forming a strong relationship between Habibullah and Smith, so much so that, eventually, Smith could walk into a village and immediately go and talk to the elders, too. Like the rest of the British Army, Smith and his officers used local interpreters, but it was good to have a United Kingdom element to it as well: while the local interpreters could be trusted to interpret correctly, there was occasionally an element of doubt, not least as their lives were always in great danger if seen to be overly fraternising.

The *shura*s would last for a couple of hours and then break up. On the whole, Smith felt that the locals of Shin Kalay didn't mistrust their motives and that most of them probably believed that ISAF were genuine. Their problem, quite simply, was that they didn't believe that the ISAF would be able to provide an adequate level of security for them.

This was a danger as Smith had no troops with which to garrison – temporarily – the village. Nor could he trust the police to run the place as not only were they simply not capable of doing so, but there were not enough of them, either.

The Afghan National Police undergo a process called Focus District Development, which involves being sent away for concentrated training. While away, they are 'back-filled' by an organisation called the Afghan National Civil Order Police (ANCOP) – a form of gendarmerie designed to maintain civil order. In this case, Smith was able to bring in about 120 ANCOP to fill the void. Generally, they were often considered to be far better trained than the local police as they understood the role of being policemen (their downside was that when asked to man a checkpoint, they would say they would rather be investigating crime).

Battle Group Centre South was to form such a good relationship with the ANCOP that there was a period when the police force in Shin Kalay was made up of – probably for the first time in recent history – relatively professional policemen. This was particularly useful as Smith needed to move C Company of the PWRR from their position outside the village to use them in his next phase, the battle for Zarghun Kalay, and he had only four days in which to effect the transfer of security of Shin Kalay from a British army infantry company to a company equivalent of the ANCOP.

With Shin Kalay secured, the elders were as happy as could be expected after being overrun by foreigners, who are often not considered any improvement on the Taliban. Nevertheless, the security and development *shura*s had reassured them to a certain extent, especially as they had been conducted under the careful chairmanship of Governor Habibullah.

With the weather starting to compete with that of a Far East monsoon and the Taliban unsure of what to expect next or where to expect it, it was vital to continue the momentum. With Shin Kalay safely under ISAF control, all thoughts turned towards Zarghun Kalay; a far more complex problem in itself.

CHAPTER FOUR

ZARGHUN KALAY

In the melee, one pressure plate device went unnoticed. As Bedford jumped across a ditch, he waited for his oppo to do the same but just as he caught him, there was a massive explosion from behind them. All he could hear was the sound of twisted metal and glass getting louder and louder as it came towards him.

The fourth Viking of the convoy had gone slightly off the track.

On 15 December, Colonel Martin Smith and L Company, having yomped up from their position halfway along Route Somerset, regrouped in Argyll to discuss plans for the securing of Zarghun Kalay. Shortly afterwards Smith issued his orders. That night L Company formally joined Battle Group Centre South and waited for their orders.

With Smith, too, was Kev Moss, his RSM-who-wasn't. Moss had been 40 Commando's regimental sergeant major during their *Herrick* tour but was now in his final two months in the Corps. Smith, however, hijacked him to fill the role of RSM, although he refused to be referred to as such. Concerned that if the Corps RSM found out he was calling himself RSM he would be 'put right in no uncertain terms', Smith assured Moss that he would be referred to simply as the Battle Group WO1. His role, however, was very much the traditional one, and included travelling everywhere with Smith, finding out what was 'really' going on at any given time among the NCOs and men, as well being responsible for the evacuation and documentation of all casualties and the issuing of ammunition of the correct types and to the correct and appropriate scales.

Zarghun Kalay was about four kilometres north-east of Argyll. It was not the usual walled village common in the region, as the south-west-facing front of the settlement was open, made up of just over a kilometre of individual compounds and houses.

As an operation, the take-over of Zarghun Kalay was expected to be far harder after the near anticlimax of Shin Kalay. It was home to many fighters from Pakistan – and even further afield – who were fighting for an ideology rather than for money. The area was certainly a staging post and a possible logistic and command node for them, and it was assumed they would fight hard for it. It was a particular area of interest for the security forces, too, as the 107mm rockets aimed at Lash PRT on 11 October had been fired from the open ground just east of the settlement.

Smith held his confirmatory orders group at Argyll on 16 December, during which he explained his deception plan prior to the assault on Zarghun Kalay.

Smith planned to put the PWRR's Reconnaissance Platoon and L Company forward, to be followed by J Company, while leaving the CO of 42 Commando – Charlie Stickland – with K Company and the Commando Reconnaissance Force to continue securing the southern checkpoints. He used the Estonians to secure Char-e-Anjir, having feinted towards Zarghun Kalay.

That night, he ordered the PWRR's Reconnaissance Platoon, in their Scimitars, to feint a recce north of Shin Kalay towards Chah-e Mirza, hoping to convince the Taliban that that was the primary direction of his advance. Once the feint had been made, Smith ordered the Recce Platoon to then provide over-watch along the final route to Zarghun Kalay.

At 0330, shortly after the Recce Platoon had left, J Company began their move out to the expected battle via a forming-up position just to the north of Luy Bagh.

Despite – perhaps because of – the low-key response to the latter stages of the securing of Shin Kalay, the Taliban had been expected to respond in some way. Sure enough, they had done just that, occupying and fortifying a line of compounds running north of Shin Kalay and Nad-e Ali, which would have to be cleared, or by-passed.

There was an added complication, too, which further showed the ISAF expectation of a warm welcome to be coming true: to prevent wheeled movement, the enemy had dug ditches across all roads and tracks that led to Zarghun Kalay from the south. Tedious and time-consuming, this was an entirely unexpected happening and meant that one light- and one medium-wheeled tractor would now have to be brought up to fill them.

The men of the battle group had known that during the approach to Zarghun Kalay they could face a 360-degree battle. This, however, would suit them as they knew that positional battles were more easily won against the Taliban who, it had been long acknowledged, were in fact at their most dangerous when they popped up unexpectedly.

Smith was also keen to use the Estonians; their mechanised infantry had a fine reputation under the command of the aptly named Major Rigor – a stern, self-contained man. Smith went to meet him to explain his plan, asked if it was within the Estonians' capabilities and wondered how Rigor felt about it all. Rigor had replied in the affirmative, with as few words as possible.

The Estonians needed no more direction than a United Kingdom infantry company, as they were highly respected for being impressively professional: once told what to do, they did it. Smith was a mortar-admirer – in the close areas of the green zone and within the loosely scattered compounds, the 60mm mortar was exceedingly useful – and the Estonians' mortars were to be particularly effective supporting L Company at Yellow 4, when the foreign mortars were firing across the battle space supporting a

British company, having had little or no cross-training about their standard operating procedures.

Once he had Zarghun Kalay, Smith planned to move straight into Chah-e Mirza in order to capture the major canal crossing at Qari Saheb Kalay – Checkpoint Yellow 4 – six kilometres to the north-east. After that, it would be important to dominate the whole of the Chah-e Anjir triangle area as a stepping stone to taking Babaji, which lay further to the north-east. Smith still had everything to play for, as the Chanjir triangle – as the British tongue pronounced it – was central to any plan. This distinctive, self-contained and troublesome triangle of small streams, ditches and dozens of isolated compounds and small buildings was an insurgents' paradise. With Chah-e Anjir at its apex in the south, just a few kilometres north-east of Zarghun Kalay, and with its base, bordering the desert, to the north, its domination would be the vital first stage for any move further north.

As is so often the case in military operations, however, if it is not the enemy that alters the plans then it will be the weather. By 0500, the Afghan weather was true to form and even worse. Heavy rain and freezing temperatures had begun to set in and with them, a low cloud base that made the helicopters, particularly the British Apache, unavailable. Because of this, many preferred the US Cobra: when the cloud base was below 3,000 feet, the Apache's sensory systems did not work as well, while the Cobra could operate below that because of different electronic systems.

On the ground and in the mire, the soldiers and marines understood the problem, but that did not allay their frustration, as the Apache had a fantastic ability to find targets then attack them with minimal collateral damage; artillery fire was far less precise. In fact, due to the very low cloud, no attack helicopters, American or British, would be available for the assault on Zarghun Kalay.

There was to be no delay, however, just because there was no air support. Following a brief discussion with the Brigade Commander, Colonel Smith decided to stick to the original plan and its timings.

By 0613, the Jesters of J Company had secured an area north of Luy Bagh without much of a fight, and established a temporary defensive position across the road and its associated ditches that ran alongside it. They were about two kilometres from Zarghun Kalay's nearest compounds and the village's interrupted line of mud walls, complete with murderholes knocked through them. Viewed through night-vision goggles and rifle sights, these were smudges of darker green that luckily were easily identified.

Smith felt able to move his HQ to J Company's position, then planned to pass L Company through. Things were made difficult by the hastily dug ditches across the roads, as not only did they need to be filled before the wheeled vehicles could go over them, but each in turn had to be checked for IEDs before anything could move. The situation was not helped either by the medium-wheeled tractor skidding off the road. The light tractor was brought up to lead but, with no armour, it could be used only once the area was fully secure. A few of the armoured vehicles did manage to negotiate the ditches but they, in turn, needed infantry support.

The situation was awkward rather than difficult, with progress still being made. But the slow rate of advance was brought to a complete halt at 0712 when someone ahead shouted that they had picked up a number of suspicious heat sources very close by. Warning shots were fired and in an instant, all forward troops were involved in a tight fire fight.

PWRR's Recce Platoon was the first into action, and moved out either side to offer support to L and J Companies by providing fire as necessary. They were also able to look north-east along either side of Zarghun Kalay to observe for possible Taliban reinforcements. The platoon was soon in a very serious fire fight, however, and they had a bogged Scimitar that was an instant magnet for enemy fire.

Everyone was in a 360-degree-fire fight with the Taliban attempting swarming tactics, desperate to get in between the companies. L Company was forward, starting to link up with Recce Company while J Company was still astride the road, close to the Tac HQ. As the battle group was being fired at from all sides – from compounds, fields, hedges, walls and ditches – Smith pushed L Company through J Company, who had gone static while returning fire, and on towards the village. It was cold, it was wet, it was confusing, and it was not helped by the fact that there were problems with radio communications, due to the heaviness of the downpour.

Smith need J Company to reduce the enemy action in the area before he could move forwards in strength and they set about that with remarkable gusto, but it was difficult as the enemy were popping up all over the place. This was not an amateur linear defence, but a real defence in depth. The enemy had carefully sited their firing ports and murderholes. They had dug trenches so they could get in and out of the area while under cover, and they were dealing with their own casualties well. The enemy was also confidently using the depth they controlled, moving defensively between compounds: left, right, backwards and forwards, while intelligence suggested that they were aiming to hold firm at all costs while conducting a significant ammunition resupply. They had obviously known for some time that the ISAF would come, and had prepared well for their arrival. Smith's men were slowly taking them out but there were lots of them: they didn't know how many.

The Task Force Helmand incident log tells the story through the briefest of statements:

0723:   Update – Six enemy killed in action in two dug-in positions. Numerous additional men positively identified with weapons seen moving in the area.

0825:   Three to six enemy engaging friendly forces with small arms. One T2 UK casualty sustained – fall from a building roof, suspected dislocated shoulder.

Lance Corporal Chris Bedford, a section second in command in 3 Troop J Company, was part of a Barma team: responsible for mine clearance, Barma teams go ahead along and to the side of tracks, looking for IEDs and similar obstacles. (Occasionally, the Barma team would catch up with the enemy, and see them laying more mines ahead. They would have to engage, and then start the searching process all over again.)

In the melee, one pressure plate device went unnoticed. As Bedford jumped across a ditch, he waited for his oppo to do the same but just as he caught him, there was a massive explosion from behind them. All he could hear was the sound of twisted metal and glass getting louder and louder as it came towards him.

The fourth Viking of the convoy had gone slightly off the track. The driver's wounds were indeed severe and as he was whisked away by the medical team the mood was sombre. Thankfully, later, the lads learned that he was 'going to be OK. However, with so much enemy fire and too much mud for the MERT Chinooks to land safely, casualty evacuation fell to the Vikings that day who, pushing forwards through all the action, were able to whisk the wounded direct to FOB Argyll and another waiting Chinook.

The advance to Zarghun Kalay was not proving to be a straight-forward task. Pinned down by sustained and accurate enemy fire on three fronts, the men of L Company, loving a good scrap, calmly identified firing points, exchanged target information and set about winning the fire fight to allow further support to cross the open ground towards the village. As RPGs and mortars continued to land around them, the troops and sections pushed on relentlessly.

Brief snapshots taken from the battle for Zarghun Kalay tell part of the story, but few capture the Somme-like mud and the crack of accurate small-arms fire, the crack of RPGs, and the indirect fire that was constantly passing at danger-close proximity. With bad weather still ruling out Apache support, L Company found itself relying on light guns and flanking armoured vehicles to make any progress.

Eventually, after a mixture of smoke, some Viking-borne manoeuvre and plenty of grit, lead elements of the company eventually broke into the outskirts of the village at about 1400. Stunned by this sudden and, to them, unexpected reversal of their fortunes, an angry insurgent counter-action followed a brief lull, and the Taliban tried to push the marines back into the countryside. L Company was having none of it, their fire support troop making tremendous use of 29 Commando Regiment's 105mm guns to quell enemy firing points within the village. The Assault Engineers were also doing a fantastic job, blowing endless holes to enable manoeuvre and clear arcs of fire.

Simultaneously, the tail of the company, out among the fire support compounds and consisting of an ANA troop, half the fire support group and the Tac HQ, were attacked from multiple firing points. However, thanks to some skilled coordination of rooftop Javelin operators and the ANA firing RPGs to keep the enemy's heads down while they moved their own light machine guns on to roofs, the position was held.

*

As night fell on this first day, so did the level of insurgent activity. It was time for Smith to consider his options; he was aware that J Company was risking becoming bogged down. Comms were still very bad; the cloud was still down so there was no attack helicopter cover; it was afternoon and starting to get dark: it was a cold, wet winter's evening. He took the decision to go firm and crack on again in the morning, hoping for a better forecast and therefore some support aircraft. Plus, if it stopped raining then that might mean having better communications.

Calling a halt to proceedings also enabled a form of reorganisation for that part of L Company that had made it into Zarghun Kalay. The situation was better than it seemed as L Company actually had its head in the village and its tail out. With no village wall as such it was difficult to say just how much ground L Company had, but they had definitely managed to infiltrate behind the enemy's front line.

Although the progress had been slow, it was not desperate, particularly considering that there had been no air cover. The marines and soldiers had faced down a determined and large enemy that had not only had the advantage of being in defence but were also in considerable depth. To cap it all, battle field communications really were appalling. Nevertheless, the aim was still to subdue Zarghun Kalay with the minimum of casualties and that would not be achieved by a continuing night attack in the appalling weather conditions, and with no air support. It made sense to go to ground and let the Taliban sweat it out.

The marines and soldiers went firm in their current positions: some in compounds but most in the open; and none with sleeping bags. Those who had twenty-four-hour ration packs shared them round. Makeshift, barely waterproof covers kept some of the rain off, while shallow channels – miniature monsoon ditches –

were dug to divert running water. Sentries were posted and support weapons manned. Apart from the incessant rain, a quiet of sorts descended across the battlefield.

The marines were all hardened to such conditions in the field by then, and although it was an Afghan winter, the temperature did not drop below about −3°C; with occasional snatches of cloud cover it could, at times, feel almost warm, despite everyone being soaked to the skin. After the tests of a long and strenuous day, some even got a very good night's sleep. Those that couldn't could at least discuss the day's events:

'How many rounds did you get in your body armour?'

'Only one, thank God.'

'I've collected two.'

'Bloody black catter!'

'Remember that bastard who was hit by a burst of 50 cal?'

'Bloody hoofing. Disappeared in a ball of red spray. Nothing left but a few fingers – and other nasty bits.'

There had been some extraordinary fighting, and there were many lucky-escape stories. Two lance corporals and a marine had been suppressing a firing point with sniper and GPMG fire when an incoming round had smashed through an ammunition tin, only to ricochet off a phosphorous grenade in one lance corporal's webbing, luckily without igniting it. Two marines had an equally unpleasant experience when a tracer round had bounced about inside their Jackal for a bit, before exiting through the rear door. And a marine and a captain who had taken cover behind a two-sandbag wall had had a narrow escpape when rounds had ripped through the top of the sandbag between their heads – and so the stories went on. Everyone collects his or her fair share of near-miss stories, but thankfully the majority return home unscathed.

*

At dawn, Smith contacted Major Rich Cantrill, the OC of L Company, to tell him that he needed him to get into the centre of the village as soon as possible. Thankfully, the cloud cover had gone, and they were about to get the attack helicopters back.

There was one troublesome compound a few metres outside the main *kalay* which, it was decided, could be silenced only from the air. With a Harrier on task, a 500-pound (227 kilogram) bomb was dropped with complete accuracy – the only one used throughout the whole operation; a fine example of economy of effort that also lessened the chances of collateral damage; an aspect always uppermost in a commander's mind.

By mid-morning on the second day of the battle, ownership of the southern half of the village by L Company was confirmed, helped considerably by J Company thrusting through on the right flank. Now everyone could help in the defeat of the enemy, in depth and conclusively. L Company's orders were to continue to secure the centre of the village while the Estonians, operating at the limits of the battleground, remained outside to prevent leakers.

Following engagements by Javelin missiles, a GMLRS guided-missile strike and some superb sniping with a .338 rifle, the insurgents began to withdraw, and the battle turned entirely in the battle group's favour. At some stage it even stopped raining.

Finally, the ground troops were joined by a pair of Cobras who, skimming the rooftops, entirely dominated the village as well as neutralising one enemy position that, ill-advisedly, had engaged one of the helicopters. To the end, the enemy had demonstrated a bravery and determination that was not always present in every operation.

As though to make up for their non-appearance in the bad weather the day before, the Apaches were in constant support for day two of the Zarghun Kalay battle, as just one sixteen-minute period shows:

0713:   Patrol came under small-arms fire, returned small-arms fire with Apache Ugly 52 in support. Fired 20 x 30mm rounds to keep enemy heads down.

0715:   Ugly 52 fired 20 x 30 mm check rounds.

0723:   Ugly 52 fired 40 x 30 mm rounds at one enemy in trench line.

0729:   Friendly forces, call sign M30 at Grid XYZ, moving north along axis of advance. Ugly 53 engaged enemy to their flank.

Within twenty-four hours of this remarkable victory, L and J Companies provided a cordon for a well-attended *shura* that was led by the District Governor. It went well, coupled with the news that all casualties were stable – thankfully, all the wounded had survived the journey to Camp Bastion.

The Helmand Task Force log is a fine testimony to the efficiency of the casualty evacuation pipe-line, and the difference in time the casevac helicopters make even when planned landing sites are too hot. Extraction and treatment within the 'golden hour' – the first hour after being wounded, within which the lives of many casualties can be saved – was the aim every time, from the initial first aid received from an oppo under fire to Bastion's hospital, usually via the stretcher bearers, then the CSMs' quad bikes and finally the MERT helicopter.

1641:   Gunshot wound. Heavy bleeding, loss of blood with low blood pressure, low heart rate and weak radial pulse. First field dressing and morphine applied.

1656:   Gunshot wound to right abdomen, life signs normal.

1732:   SITREP Unable to secure helicopter landing site due to enemy fire. Intention is to send two casualties to Argyll for collection by MERT Chinook.

1741: MERT launch authority given.

1748: MERT launch authority now withheld.

1749: Helicopter landing site at Argyll Barma-ed and secure.

1806: Two casualties now at Argyll.

1808: MERT wheels up Bastion to Argyll.

1818: MERT wheels down and wheels up at Argyll.

1825: Wheels down at Bastion.

1830: Two casualties at Bastion medical centre.

Before each set-piece of an operation, battle commanders assess the casualty estimate, so that the medical back-up teams can judge if they have the correct medical plan in place. Smith was clear where he stood in this matter; he was uncomfortable with the whole idea of keeping casualty estimates but back in Task Force HQ they were running them for a very good reason – in order to aid evacuation procedures and to monitor combat effectiveness – so he had to keep track. Looking at *Sond Chara* overall, he had few casualties, and the operation came in well under the estimates.

It was assessed that Smith and his men had killed about thirty-eight Taliban at Shin Kalay. However, these estimates were more difficult to make for Zarghun Kalay because although it was a longer and bloodier battle, no bodies were ever found: by this point, it was assumed, the Taliban were much more organised, and had evacuation chains as well as medical facilities set up in compounds. The suspicion was that during the battle for Zarghun Kalay, they moved their casualties into the Chah-e Anjir triangle where their influence was strongest, and where they had rehabilitation areas; yet more proof that the friendly forces were up against foreign, well-trained fighters.

The men of Battle Group Centre South had fought with calm grit and fortitude during the battle for Zarghun Kalay, putting to flight a large, determined and professional enemy force from good defensive positions. This had been achieved in the fine, old-fashioned way, through well-led, sensible infantry soldiering. It was, by all accounts, also a psychological defeat: the resistance melted away, allowing the locals to come out of hiding and meet the marines.

Smith now did exactly as he had done at Shin Kalay, and held two *shura*s with Habibullah, the Chief of Police and the CO of the *kandak*.

The Taliban had not been destroyed in Zaghun Kalay; they had merely been defeated, before melting back into the countryside about one kilometre north of the village. There, Martin Smith knew, they would be regrouping and licking their considerable wounds.

With this in mind, Smith and his team were already planning the attack on Chah-e Mirza, about five kilometres to the north-west, where there were about five concentrations of dispersed compounds – small villages – known to harbour the enemy: an enemy moreover that was able to control the adjacent canal crossings into the *dasht*.

While the planning process was set in motion, Martin Smith ordered his Recce Platoon to carry out a deception probe to the west of Zaghun Kalay, as he was keen to have the Taliban thinking that that would be the direction of his next operation. He was, in fact, planning a left hook through Shin Kalay before swinging north to Chah-e Mirza. Again, though, this feint would have to be carefully undertaken, and Barma-ed all the way: other than the military ISTAR, Smith's HQ had received much local hearsay suggesting a vast increase of IEDs in the area.

*

If the battle for Shin Kalay was not exactly a disappointment, then the non-battle for Chah-e Mirza most certainly was. On 23 December, the battle group conducted what was, essentially, a 350-man fighting patrol, with everyone yomping overnight in what must have been the largest fighting patrol ever conducted in Helmand.

The three companies, plus the battle group's Tac HQ, moved in tactical bounds, with as-quiet-as-possible Scimitars behind, on call. Closing on their destination, C Company of the PWRR was despatched further ahead into a forming-up position four or five kilometres north of Shin Kalay, in order to surround and observe the area while it was still dark. L Company was ordered to do the same before the Scimitars would be called forwards the next day.

It was a cold night and everyone was wet. Then it became frosty. There were no cold-weather casualties, however, which was put down to the rather dubious explanation that the boys were very experienced at living in the field by now: they had been in it for weeks. But one bonus was that there was no enemy either. Anywhere.

Martin Smith thought this was odd as ISTAR believed Chah-e Mirza to be an important enemy logistical area close to a major canal crossing point. It was also an area in which the Taliban had recently destroyed every school that had been the subject of western influence. One of the most popular agreements at all *shuras* was the firm commitment to rebuild schools, but while the Taliban claimed not to object to either schools or clinics – for males – being built, they objected to anything that had been built or paid for by western nations. However, as there really was no enemy about, it was decided that everyone would wait for daylight, when they could meet the Chah-e Mirza locals.

*

The next morning the marines and soldiers found themselves entering a much more well-to-do area than anywhere else they had been of late, with the inhabitants looking well fed and better organised.

The battle group's Tac HQ was quickly set up in a field.

The lack of enemy forces was not to last long, however, and Christmas Eve was not to be respected:

0626: Fighting-age male spotted in a compound armed with a pistol, engaged by the patrol and the male was KIA.

During this very brief engagement, a ricochet hit a one-year-old child in the upper right leg, but within the hour he had been lifted from the secure landing site at Nad-e Ali and was on Camp Bastion's operating table.

A more typical incident also happened mid-morning that Christmas Eve. Two fighting-age males were suddenly seen driving erratically towards one of Battle Group Centre South's positions.

A warning to stop was shouted, but to no avail.

'Fire a burst overhead – this bugger means business.'

Three bursts of GPMG rounds were loosed off. No reaction.

'Try the engine block.'

'Roger.'

Thee more busts followed, but still the car kept coming.

'Bloody hell, mate. Go for it.'

The third burst shattered the windscreen and the vehicle slewed to a standstill.

Silence.

Colonel Smith needed to move to the major canal crossing of Qari Saheb Kalay at Checkpoint Yellow 4 in order to assist the commanding officer of 42 Commando – Lieutenant Colonel Charlie Stickland – in getting his men into the Chah-e Anjir triangle; the gateway to Babaji. A further move into Babaji was not to be a Battle Group Centre South operation, however; that much had been agreed. As it would have left the Brigade with too much ground to take and keep control of, the area was to be left for 19 Light Brigade and *Herrick 10*.

It was necessary for Battle Group Centre South and 42 Commando to meet at Yellow 4 as from there everyone could re-group, return to their parent units, or disperse to man the checkpoints that the battle group had been leaving in its wake across Nad-e Ali. It was also a chance for the marines of L Company to stop yomping. They had covered over sixty kilometres since the operation's start; during which time they had also fought more than their share of battles and even more skirmishes and scraps. It had been a quite remarkable achievement, but their work was not yet done.

The final route to Yellow 4 had not been without its problems thanks to IEDs, the general lack of which in this area had been a notable factor up to this point. Now, facing a significant threat, Smith could either bring in the Barma team to deal with them, or try and find another route to save time. However, intelligence confirmed that all the routes in the area were probably mined as well as being heavy with Taliban, so there was nothing for it but to forge on slowly and carefully, with Barma teams out in front.

Eventually they reached Yellow 4 and L Company were once again embroiled in fierce fighting. Finally they linked up with 42 Commando's Tac HQ and CRF would have flown into the *dasht* and yomped in to meet them. Smith had problems with vehicles, and his Tac HQ Viking had had to go back down the levee to tow in another vehicle. As a result, Charlie Stickland, CO of 42, and he, could talk only on the radio, and weren't able to meet to discuss the situation.

Smith then moved out to start occupying all the new checkpoints he had put in place. He gave L Company back to 42 Commando – they could take a deserved rest at Yellow 4 – and Charlie Stickland moved in to start concentrating on the push for Chah-e Anjir triangle the next day.

Smith now folded back down the levee to Chah-e Mirza, where he could begin rejigging his forces. It had always been his intention to bring C Company of PWRR into Argyll so that he could return to his main HQ back to Bastion, and this he did. C Company became the Nad-e Ali Company, J Company returned to Lashkar Gah, while the Estonian Company moved to Nawa.

Movement was becoming more and more constrained as the IED threat increased. The Taliban may have been defeated in battle and the main urban centres cleared of them, temporarily, but in the wide spaces that made up much of the Nad-e Ali district, the Taliban were still very much an extant force, with IEDs now their predominant weapon.

The fact that the Taliban had offered remarkably little resistance in recent action when compared with what they were capable of achieving during, for instance, the push for Zarghun Kalay, was in no doubt. In the end, it appeared that they had simply decided that it was in their better interests not to be seen to be operating against the population of Afghanistan. Despite some obvious evidence to the contrary, the Taliban did not want to disrupt the locals too much: fighting through their villages and compounds would not endear them to the people on whom they relied for support – freely given or otherwise. Their increased use of IEDs was, therefore, a good sign for the British: coupled with the Taliban's reluctance to alienate the local people through their use, they must have been concerned about the state of play if they were now falling back on using them. There was also a reluctance to use IEDs as it was considered a sign of defeat and, for a Pashtun warrior, it was a cowardly act to use them: they needed to be seen to be people who could take on the enemy face-to-face and win. This increased use of IEDs could mean only one thing – the Taliban were worried.

There had to be an interlude to such non-stop kinetic activity. There was, of course, no alcohol and in FOB Argyll, there was no turkey for dinner, either. However, there was a goat and so this unsuspecting animal, before being led to the slaughterhouse of the Royal Marine chefs' kitchen, was swiftly renamed 'Turkey': although bad luck for the goat, it did at least mean that every soldier and marine could write home and say that, despite all other privations, he had been able to eat turkey on Christmas Day.

For others though, Christmas Day was just like any other. Early on Christmas morning, a helicopter assault by Stickland's 42 Commando was launched against an infamous safe haven for the enemy in the Chah-e Anjir triangle. The ferocious assault significantly disrupted the enemy in this strategically important area. One of the objectives taken was identified as a key enemy command and control and logistics node, and was linked to the attacks on Lashkar Gah on 11 October. After an initially fierce break-in battle, all manner of Taliban equipment was discovered.

Meanwhile, for those left behind at Patrol Base Stella over Christmas, it was an uncomfortable existence under daily contact from rogue elements of the Taliban that had been dislodged from their *kalay*s. Gradually, though, the enemy retreated out of range of the Manoeuvre Support Group's optics and weapon systems, and the area quietened.

Plans had been made to build on the Nad-e Ali success with a similar series of operations into Babaji and further up the Helmand Valley, but apart from a disruption helicopter-borne raid there were not the resources of men and *materiel* to do more. The planners, though, passed on their ideas to 19 Light Brigade that, with extra manpower, was able to execute *Operation Panther's Claw* in June 2009, after the 3rd Commando Brigade had left. One advantage for 19 Light Brigade would be the resource of 1,000 ANA soldiers holding the line of departure when the Marine Brigade left, so that they would be able launch straight into, then clear through, Babaji.

With a successful Battle Group Centre South dispersed to checkpoints – after having been formed only eight weeks before – Smith now spent a whole week with Habibullah, the Chief of Police and the CO of the *kandak*, trying to get everything on an even keel prior to voter registration. All was now set, as well, for the distribution of wheat seed and the poppy eradication programme – both inextricably linked.

Without Operation Sond Chara, the secure conditions needed for voter registration would not have been set, but it is necessary to acknowledge that without Provincial Governor Mangal taking advantage of those conditions, there would have been no voter registration at all.

As can all too easily happen, Mangal could have wasted the opportunity offered to him by ISAF, yet he was to get it absolutely right. By doing so, in early 2009, every eligible male and female was able to walk in to designated compounds, have their fingers inked and then their names checked against the electoral roll. When the time came, all those who had registered were able to vote for the President of their choice.

With *Sond Chara* completed, the Governor was able to stand in the middle of Nad-e Ali – which he could not have done before without risking being summarily gunned down – and address the largest gathering of people – about 2,000 – seen there since the Taliban had become powerful. He introduced the recently appointed district governor with the words, 'Here is Habibullah, who will allow you to have local elections. Today we shall have a *shura* and decide who you want to be on your town council. Money will come from Kabul through me and we will start build-ing schools [two were completed by the time the Brigade left] and we shall build a police station and distribute wheat seed for you to grow and sell.'

For Governor Mangal, once regarded as a dyed-in-the-wool communist, it was a great day. Considered a reasonable chap, he was also regarded as being good news following some awful provincial governors. On arrival in 2008, he had soon been accepted as an effi-cient and effective man and very much one from whom to expect great things. By Afghan standards he was squeaky clean, and was always interested in the PRTs' work and, through that organisation, the work of the military; its failures and its successes. He even came to be regarded as part of the information campaign and was so trusted that he would often be given sensitive information about missions prior to their start: he would be asked only to release infor-mation to the police on a given day so that the PRT and ISAF could prosecute a particular operation smoothly.

Voter registration and wheat seed distribution were interesting problems and ones that the Taliban didn't wholly understand. The success of the former was evident when the 1,000 people that were expected to apply for registration turned out to be 60,000, out of a possible 185,000. In the face of such popular support for it, there was not much the Taliban could do other than in tiny, away-from-the-public-eye incidents: if a member of the Taliban found someone with a registration card, for example, they would force them to eat it. The reality was that the people of Afghanistan voted for whoever would, or could, deliver security. It was all they wanted – security.

Poppy eradication was a concern in which Mangal led the way, by issuing 100 tractors to local farmers for one specific reason. Calling on local villages, the Governor issued maps taken from satellite images to show where he expected action, telling the farmers that there were no excuses, and that they were to start ploughing their fields and sowing wheat. The original target for success was low, but in the end he was to achieve a 90 per cent success rate; he knew that there were probably huge backhanders being paid by the Taliban to some farmers not to plough certain fields.

He ran this alternative livelihood programme through the Lash PRT, so it was very much a civilian-led business, which also included the wheat planting. Once seed had been delivered, he would then tour the area in a helicopter to check that wheat had been planted. Where wheat seed had been delivered, but poppies rather than wheat was being grown, Mangal would confront the farmer, and was protected by law in ploughing in the opium poppies and destroying the crop.

He was also able to guarantee a fixed price for the wheat grown, and the PRT then organised the delivery of about 95 per cent of the wheat in and around the locality. This delivery formed something of an aid delivery, with even a cheeky bunch of Marjah Taliban turning up to be issued their free wheat – they had to do something, as they were almost powerless to affect the wheat issue programme, which was heavily protected. The important factor was that the eradication was a Governor-led programme, rather than a NATO one.

The bigger challenge was poppy eradication, with two elements making up the force combating it: a private military security company called DynCorp provided leadership, and was supported by a counter-narcotics infantry *kandak* (CNIK). Both were paid for by the US State Department. DynCorp undertook all the farming duties once they spotted poppies; ploughing and replanting the fields with wheat seed, while the CNIK provided protection. The practicalities of this task were immense, as it was not always physically easy to reach the fields, many of which were bounded by irrigation ditches and linked by makeshift bridges over which the tractors could not pass. An added complication would occur if a farmer had planted wheat but knew that compensation for destroyed crops was greater than the market price.

The key to much of this progress were the stability advisers. Jim Haggerty and his fellow STABADs worked hard in conjunction with Habibullah, authorising him money and the ability to make projects happen, usually in conjunction with the PRT.

In Nad-e Ali, and following Martin Smith's battle group's *Operation Sond Chara*, a partnership between the military and the PRT was being forged that had never existed before. Thanks to their efforts, there were now two groups of people of totally different backgrounds who had to operate together, and who were now operating together. It worked well, and there was mutual respect and accord on both sides.

There was, finally, the beginning of a structure in the governance of Nad-e Ali and, in fact, the most satisfying of all *Herrick 9*'s experiences would prove to be the close relationship between Task Force Helmand, the PRT and the local governors: a relationship that was very much to the advantage of both ISAF and the locals. Many considered it a great pity that it had not happened before.

*

Martin Smith and his men of Battle Group Centre South had defeated the enemy in Nad-e Ali and given control of the area back to the Government of Afghanistan. *Sond Chara* had achieved its aims, and also had come in below the casualty estimate. Although there was some frustration that there wasn't the manpower to be able to finish the job and proceed to Babaji, the newly established battle group had also laid the very firm foundations needed for the springboard from which 19 Light Brigade could launch *Operation Panther's Claw.*

CHAPTER FIVE

SANGIN VALLEY

*Shortly after midday, the patrol was caught in an immense explosion.
The leading section commander was killed instantly, while in front of
him, the point marine was blown across a ditch into an open field,
very seriously wounded. The third man was 'Bugsy' Malone who, still
clutching his GPMG, was hurled off his feet backwards. Shaken and
bruised but otherwise OK, he collected his wits, shouted for two men to
come forward to help and then grabbed a radio set. The Taliban had
other ideas, however, as before the dust settled the whole patrol came
under fire from four well-prepared firing points: the closest was 100
metres to their right, with another two beyond that and the fourth
200 metres to the left. It was a classic IED-initiated ambush.*

During *Operation Herrick 5*, 45 Commando had the interesting and vital task of mentoring the Afghan National Army. Now, for *Herrick 9* and under the command of Lieutenant Colonel Jim Morris, 45 Commando was selected for the business of relieving the 2nd Battalion the Parachute Regiment in Sangin as Battle Group North.

While they were to take on what were traditionally the bad lands of the Upper Sangin Valley – as far north as the Kajaki dam – Battle Group North's overall mission was to secure Sangin DC and Kajaki and disrupt enemy forces in the Upper Sangin Valley while, in partnership with the PRT, deepening the Afghan Government's control in Helmand.

During the course of their tour of duty, 45 would establish the daily task of a non-stop patrolling programme operating out of their forward operating bases: hard, footslogging, old-fashioned soldiering. *Operation Diesel* – as part of the *Aabi Toorah* series – in the Upper Sangin Valley, was the only exception. For that operation, Jim Morris was given all the combat power he needed and told to get stuck into the IED factories of Sapwan Kala further to the north, as most of the events that happened in Sangin had their roots in this area.

Jim Morris joined the Royal Marines in 1987. He had served previously in the Commando in various regimental appointments. He had also served in Afghanistan in HQ ISAF, Kabul as Chief CJ3 (Operations) during the time that the Brigade was conducting *Operation Herrick 5*. One of the benefits from these tours was that he knew most if not all the other commanding officers, as well as the Brigade staff – something that doesn't often happen in an army brigade, as those who make up the component parts will have come from across a large army, whereas those in the Commando Briagde have known each other and worked with each other throughout their careers.

Beyond 45, support was equally as good. Nick Pounds was Jim Morris's stabilisation adviser. Although Pounds had been a Royal Marine brigadier, the two had never met but the relationship developed quickly. Along with Pounds, and key to much that was achieved, was the Governor of Sangin, Haji Faisal Haq, who had relieved a corrupt predecessor. Uneducated, illiterate and unsophisticated, Haq was not an ideal politician but he was politically astute, increasingly popular and very much the right man for the moment. Haq came from the Sarevan Qaleh area north of Sangin, was anti-Taliban – he had raised a *levée en masse* against them – and had fought the Russians. Brought in by Gulab Mangal, he was the first Governor for some time to have a local background and relatives in the area.

Morris soon got cracking. With Nick Pounds and their respective planning teams, he designed a comprehensive plan across the whole of his tactical area of responsibility. Together they devised the Sangin Security and Stabilisation Plan, which defined what was needed in security terms to deliver Pounds's desire to create a Government Zone – an economic zone – to push forward governance in Sangin. The District Governor and the Commanders of the Afghan National Army Kandak and the Afghan National Police were involved throughout this planning. It was seen as an Afghan solution to an Afghan problem, but to achieve it they had to get on with establishing security, too. Alongside this, voter registration was all part of extending governance, as was the wheat seed distribution project. There was a lot to be done.

Morris's men of Battle Group North were in fixed locations and patrolled out daily. He was reluctantly pragmatic about the inevitable lack of collective equipment; he had just enough helicopters to do what he was asked to do, although for day-to-day tasks he had to bid for them long in advance, which made putting the right plan together to get the right resources crucial. The battle group did not, however, have a huge allocation of vehicles – only enough for casevac and manoeuvre support. They did have Jackals – a few – that proved to be the most effective vehicles for the work they did.

The transfer of authority took place on 14 October. The 2nd Battalion the Parachute Regiment had had a hard time, but gave 45 Commando a perfect handover, making sure that they had everything they needed at staff level and on the ground: the operations were recognised by all to be far bigger than any cap-badge rivalry or politics. Morris was to make sure that he gave their successors, 2 Rifles, similar honours.

On 45 Commando's arrival in Helmand, it was clear that a significant shift of emphasis was taking place. Task Force Helmand's main effort was moving away from its concentration on the Sangin Valley to focus on the threat to Lashkar Gah in the south, and the resulting formation of Battle Group Centre South had led to the departure of many of the assets that 45 Commando and its out-stations relied upon for its eyes and ears. Simultaneously, the United States Marine Corps presence in Sangin was redeployed elsewhere. As the security of Sangin was the *raison d'être* for the Commando's presence in that part of Helmand, Morris would have to make good these shortfalls from within his own resources.

Part of his initial task, therefore, was to re-balance the Commando to meet their commitments. Added to this, Morris was keen that the tour of duty was not just going to be a 45 Commando six-month show. He was keen to build on the progress of his predecessors and to hand over something positive and concrete for his successors to build upon.

That said, the rebalancing of the Commando did have an impact. For example, Y Company had to send one of its Rifle Troops to Sangin to help plug the manpower shortfall, meaning that they were cut to two-thirds of their combat power. They were to remain like that for the rest of the tour. The gunners, engineers and various other vital attachments stayed, but this reduced force (50 per cent of their predecessor's strength) still faced an unchanged mission: that of intercepting and disrupting the Taliban to the north of Sangin to reduce their impact in the Sangin district centre.

This was a daunting task. 45 Commando inherited a yearly troops in contact rate of 281 and an IED rate (found, and set off) of 280; by the time they handed over there had been, respectively, a 22 and 94 per cent increase (*Herrick 10* was to see yet further increases). Coupled with this, any foray into the green zone beyond 300 metres attracted a scrap, yet patrolling was a necessary function for reassuring the population, dominating the ground and keeping the insurgents in check.

It was not the fighting that was to be a worry for the marines, as they were confident in their ability to win: it was the lack of manpower, more specifically when faced with conducting an extraction of a casualty while still in contact. In order to moderate this, Y Company patrols into the green zone became an all-hands-to-the-pump affair, with marines being pulled from their daily specialist roles to man patrols to keep numbers up. It was skeleton-crew time. Never was it more important that all Royal Marines were trained as commandos first before they were trained in any other specialisation.

In real terms, the Commando had deployed into Helmand with an all-up strength of 707 personnel: ninety fewer than 2 Para and yet still with the same tasks. Added to this shortfall was the extra complication that some of those duties had been undertaken by 235 men of E Company of the United State's Marine Corps' 24 Marine Expeditionary Unit, something that came to light only four days before the transfer of authority. Colonel Morris was suddenly aware that he would be, effectively, 325 men short.

Another absentee was the cultural adviser, but this loss was partly compensated for by a very good local interpreter who had spent some time in the UK. He was a brilliant person for the job, and was utterly trusted in the HQ.

Commando HQ was co-located with the Governor in Sangin district centre, making it easy for Morris speak to him throughout the day, attend the Governor's meetings and to join him in the evenings, and catch up on the day's events. Occasionally, to make a point, the Governor would serve what he described as a 'poor people's meal' – Afghan style.

Morris had arrived in Sangin a few days earlier than the transfer of authority as he was keen to sort out the problems thrown up by the sudden shortfall of manpower.

45's Companies staged through Camp Bastion individually for their RSOI (orientation and integration training), which was conducted by teams from the outgoing troops. Despite some believing that these duties would have been best carried out by specialist and permanent teams, it made complete sense for it to be carried out by those fresh from the field. No matter how many times the lads had been to Afghanistan, it was worth the time, as the situation changed so rapidly.

Morris had five manoeuvre companies under his command for Battle Group North: the 45 Commando regulars were W, X, Y and Z Companies; and V Company, which was formed specifically for *Herrick 9* and was made up from men in the marines' home base at Arbroath on the east Scottish coast.

Once individual RSOIs had been completed, the companies could move to their forward operating bases where, in each case, they were to conduct their relief in place while under fire. Despite this, and probably because of this, every marine declared himself happy with his handover.

Being split between five forward operating bases meant that the battle group's different companies each had their own unique set of tasks and challenges. In the north, V Company in FOB Zeebrugge was tasked with protecting the strategically important Kajaki dam. In the centre of the battle group's area of operations was the crucial area of Sangin district, which was carved up into separate areas secured by Y Company, operating from FOB Inkerman; W Company from FOB Jackson; and X Company from FOB Robinson and then FOB Nolay. To the south, Z Company monitored and intercepted enemy forces moving between Gereshk and Sangin, and operated out of FOB Gibraltar.

While Colonel Morris had to ensure that each of his company forward operating bases was appropriately manned his priority lay in Sangin. The need to rebalance the Commando meant that he had to look closely at Kajaki and at FOB Gibraltar – already regarded as an economy of force operation – then boil down his resources where he could, to shore up the gap left by the US Marines' redeployment.

Battle Group North's mission was varied, but always based on the principle of 'clear and hold'. The task that faced it was to clear and hold Sangin, plus all the surrounding and supporting forward operating bases, including the Kajaki dam.

The CO recognised the strategic importance of the Kajaki dam and understood that his men were providing vital protection to the United States Agency for International Aid project, but felt they could be better used elsewhere with others providing protection for the dam; FOB Gibraltar to the south of it was, in fact, of greater tactical benefit to Gereshk than it was to Sangin and FOB Robinson, six miles south of Sangin, was a legacy patrol base that also housed members of the US forces unit employed in training Afghans, all of whom relied on the resident British infantry company for their defence. 40 Commando, a year before, had recommended moving the base closer to Sangin, as had the Parachute Regiment more recently.

Now, as part of his rebalancing of Battle Group North, Morris recommended the company in FOB Robinson be moved, with its combat power, to the north to cover Sangin where it could dominate the southern approaches to the town much more easily. Not only was this a good idea from a rebalancing point of view, but from a tactical perspective it removed the need to patrol through a well-known IED 'belt' which was rather like a minefield and which would take up to four to five hours each time to move through. The powers that be were initially surprised when Morris made the case for a move further north by only a few kilometres, but the location and the realities of navigating an IED belt on a daily basis just to conduct a dominating patrol meant that anywhere else had to be better.

*

The new FOB – named Nolay – was far better sited for dominating the ground south of Sangin. Its mission was to protect Sangin from enemy approaching from the south and through the dry wadis to the east. From it, the troops from X Company patrolled both the green zone and the desert fringes, regularly disrupting enemy forces on their approach routes. During these patrols over the months, they found caches of weapons, ammunition and explosives that the enemy had concealed for later use in Sangin.

Once the base was properly established, it also started to produce results by giving the people greater protection and thus a greater sense of security. With people feeling safer, trade returned, as did the schools, and the finds of IEDs, weapons and ammunition stopped the Taliban using the area as a staging area for attacks and pushed them out of the south of Sangin.

X Company's leaving of FOB Robinson and the establishment of the new FOB Nolay was, as expected, far from incident-free, and Major Rich Maltby was the company commander upon whom the complicated and delicate task of conducting the move fell. As he had walked from the helicopter pad to the HQ building on his arrival at Forward Operating Base Robinson on 13 October 2008, he had known, in that short distance, that to stay in FOB 'Rob' for one minute longer than necessary would be a mistake. It was a shoddy place and in full view of the enemy, yet with no ability to cover the approaches to Sangin. The enemy could move up the far side of the river or they could come in from the desert; either way, by-passing the base. To be of any tactical value, it needed to be closer to the district centre.

Robinson, with its legacy status, had long outlived its usefulness. It was in the wrong position, could be over-watched by the Taliban and was easily surrounded by IEDs. Coupled with this, the FOB was also suffering from an identity crisis: no one seemed to know whether it was a logistics node, a patrol base or even a fire-base. It was, though, unanimously agreed that it was unsuitable as a commando company location.

The Paras had identified a collection of compounds, two kilo-metres to the north, closer to Sangin, and thus in a position more able to influence the movement of insurgents. Comparatively smart and desirable, the proposed new compounds, which were oddly unoccupied, had been nicknamed Millionaires Row.

Maltby decided to organise a patrol to assess the likely new loca-tion. The reports were encouraging. Situated on an escarpment, the proposed compounds overlooked ground almost up to the Sangin district centre itself, and were a stone's throw away from the local population that X Company was hoping to influence and protect. With far greater arcs of fire over a significant section of the countryside, the decision was an easy one to make.

Outline plans were drawn up and debated. Once these had been decided upon, they were put to Colonel Morris. After consultation with the Brigadier Messenger and subsequently the Chief of Joint Operations at the Northwood-based Permanent Joint Headquarters, Morris passed their thumbs-up back to Maltby. The major now had a tight deadline to work towards, with his ambi-tious relocation plan facing a number of difficulties.

What looked to be a straightforward move – by Helmand stan-dards – from one location to another just a few kilometres away, was to test the men of X Company severely, particularly as any move had to be conducted swiftly so that the company would remain at peak fighting capacity at all times.

The dual problems were a tenacious enemy and bad timing. The preamble included moving a great deal of Royal Engineers' stores, and that needed the certainty that the one Coles crane would be serviceable and ready to assist in moving eighty-five laden containers. A bulldozer was also needed, as were quantities of explosives and detonators. Importantly, the residents of an American task force, who would remain in FOB Robinson, needed advance notice that they were about to be responsible for their own security and defence.

With as much in place as was possible, and after six weeks of planning and constant patrolling including intensive and successful negotiations with the landowners, the day selected for the start of the move was 19 November. Early in the morning, the Royal Engineers left FOB Robinson, escorted by an assault troop of marines. Their destination was the higher ground beneath the walls of the left-hand compound of Millionaires Row.

Tasked with preparing the mouseholing charges for entry to the compounds, Corporal Ross Austen conducted the Barma, on foot and ahead of his team, until they were close to the escarpment beneath the compounds. He then crept ahead to place two mouse-hole charges; the marines were taking no chances and were going to enter the compound by the quickest route they knew.

With his charges in place, Austen began the return journey across the short distance to safety; once the holes were blown the marines would rush through and the occupation could begin.

But the explosion his colleagues saw was not against the compound wall; instead, one blew from the ground, enveloping Austen in flames, smoke and dust. While some rushed to his aid, others radioed for help. Both of his legs had taken much of the blast and while they were not to be amputated, they lost most of their muscle. Sadly one leg was later to be amputated.

1233:   One category B, UK Military casualty brought to FOB Robinson.

1246:   SITREP Engineers, who were blowing a compound wall, were running back to troop location when they were involved in a contact explosion.

Thirteen minutes later, the MERT helicopter was airborne from Bastion. It landed at FOB Robinson at 1311; by 1334 Corporal Austen was back at Camp Bastion under the care of the medical team.

Unable to use the main road as it was almost permanently mined – one improvised mine had been formed around an unexploded 500-pound ISAF bomb the Taliban had found – Maltby's men were obliged to forge their own safe route between the two bases. In honour of the Royal Engineer corporal, it was named Route Austen.

Despite this setback, X Company secured the compound without further ado, apart from the fact that in the next-door compound – earmarked as the HQ – a wedding party, Afghan style, of over 300 men, women and children, had secretly moved in and was in full swing. This was clearly a delicate situation, with the potential of becoming a mammoth public-relations disaster. However, clear thinking and some fast negotiation by Major Maltby found an agreeable compromise: in his resulting speech, he thanked everyone for coming, complimented the bride on her beauty (gallantry itself as he couldn't see her beneath her burka), and informed the guests they had twenty-four hours to finish the festivities. His negotiations worked, and by the following evening all the occupants and their guests had left.

The difficult task of fortifying FOB Nolay, while patrolling, now lay ahead, as well as the task of shifting between the bases the eighty-five containers, two of which were full of 105mm ammunition – not that the FOB had these guns. Other such anomalies were steadily dealt with over the weeks, during which the sappers worked longer hours than most, while men of all ranks became proficient at filling sandbags.

The establishment of FOB Nolay made a huge contribution to the success of Battle Group North's tour of duty. Patrols not only reduced the Taliban's freedom of movement towards and away from the district centre, but they were able to visit areas previously unvisited by ISAF forces.

Disaster was to come X Company's way again, however. On 12 December, a patrol was guarding a small bridge over the canal a kilometre to the north of Nolay, near the small village of Pan Kelah Shomali. It was in the heart of Taliban country, but over the weeks since establishing the forward operating base it had been hoped that ISAF influence had been pushing them slowly back.

The men of the patrol were out of their vehicles, observing, and watching the area carefully. At the same time a young lad in his early teens was pushing a wheelbarrow laden with papers towards them; in those rural areas, full wheelbarrows were a common sight.

The explosion was massive and deadly. Sergeant John Manuel and Corporal Marc Birch were killed instantly. Marine Damian Davies was critically injured, and a large number of innocent civilians were seriously wounded.

The terrifying cry of 'Man down' was once more yelled down the radio by the patrol's radio operator, followed by 'Wait out!' After a lightning-quick assessment of the situation, he then began calmly passing on the details to set in train the emergency evacuation procedures.

1051:   Contact explosion resulting in 2 x T1 casualties and 1 x KIA.

1110:   UPDATE Contact explosion at Nolay was suicide IED.

The on-call helicopter had already been scrambled and was airborne at 1113.

> 1122:   UPDATE Suicide IED was a young boy pushing a
>   wheelbarrow with newspapers covering the device.
>   Second T1 is now confirmed KIA.

Twenty minutes later the Chinook landed at Nolay's recently constructed helicopter pad, within the compounds. It stayed for only the two minutes it took to move Damian Davis up the stern ramp before taking off for Camp Bastion.

Tragically Marine Davies, on attachment from the Commando Logistic Regiment, did not make it either.

It was one of Battle Group North's darker days, as Lance Corporal Jamie Fellows was also killed in action as he patrolled out of FOB Jackson in Sangin district centre earlier that day. X Company also had a T3 casualty reported shortly into the afternoon, which also needed emergency evacuation.

It was to prove a busy time for the Chinooks across the Helmand Task Force's area of responsibility, as at the same time *Operation Sond Chara* was in full swing in the Nad-e Ali district.

Operation Silver, during the very last days of Herrick 5, had effectively cleared Sangin of Taliban, but it was always difficult to hold somewhere – especially somewhere as complicated as that town – once it was cleared. Nevertheless, people had begun to return in 2 Para's time, with reopening businesses developing the whole area into a serious economic zone. By the time 45 Commando left in 2009, the town was thriving once more.

Morris's interest in Sangin did not go unnoticed by the Taliban, who started to fight back fiercely: the battle group was taking them on in their own backyard, and they didn't like it.

From the very beginning of their tour of duty, the IED rate had increased daily. Morris had realised quickly that he had to do something about it, because it was restricting the freedom of the battle group's movements, while also preventing the civilians from going about their business in Sangin. They could not allow the Taliban to dominate them, nor the inhabitants of Sangin to leave again.

Morris started to introduce some forward-leaning operations in conjunction with the Task Force's counter-IED personnel. They knew the ground that the Taliban held and so knew that they would resow the IEDs each time friendly forces lifted them. The plan was, then, instead of engaging and killing them as they did so – something that would have been perfectly legal – to exercise a little tactical patience. This meant, coupled with ISTAR assets, they were able to track the Taliban back to their various places of manufacture and storage, which gave them a whole range of target sets, and the company commanders were able to exploit their ability to be proactive and fight the IED menace at its source.

Before deployment, Morris had earmarked W Company to look after the heart of Sangin, and things at FOB Jackson, on Sangin's western outskirts, were busy. The company had started with 108 ranks, which in the space of two months, due to Morris's rebalancing, more than doubled to 225. With these numbers came the responsibility of manning three patrol bases: Wishtan, Tangiers and Pylae. The first two locations were at troop strength (approximately thirty men), while Pylae was at section strength (around ten men), and mentoring a group of ANA, which created its own challenges.

The patrol bases had to be resupplied by road; a nightmare as this meant creating patterns of behaviour, and therefore exposing themselves to the increasing IED threat. On average, FOB Jackson's patrols were finding and destroying one IED every thirty hours and, over their tour, they were to have thirty-five IED strikes, causing some horrific casualties.

The first was Sergeant Paul 'Baz' Barrett, a keen company mountain leader. He stood on a pressure-pad IED made from over thirty kilograms of homemade explosive. It took off his right leg and a large portion of his right arm. To complicate matters, as the strike had happened in a wadi, the casualty evacuation involved having to move him from a stretcher to a quad bike, which constantly sank into the wadi's sand. They then had to transfer him into a battlefield ambulance and eventually on to a helicopter. All of this, however, was achieved in just over forty minutes from the time of initial contact, which was huge testimony to well-carried-out casevac skills.

W Company found this incident particularly hard as Baz Barrett was a very popular man. Along with this, conditions in Jackson were far from ideal as, with the approach of cold weather, the FOB became a building site, with engineer plants constantly moving round the forward operating base preparing it for the bite of the Afghan winter.

The majority of the men were housed in Fire Support Group Tower, which resembled an ancient fortress and was strewn with heavy weapons on the roof to provide 360-degree protection. The inside was even more medieval, with no heating whatsoever and the marines huddled round candles, wrapped in sleeping bags, woolly hats and gloves to keep warm, while the few showers for over 400 ranks were freezing at best. Eventually, they were provided with two forty-inch plasma televisions with a DVD player, as well as being able to tune in to British Forces Broadcasting Service, so morale was raised considerably.

*

The Patrol Base in Wishtan was, even by Afghan standards, an austere place with the thirty or so marines drawn from Zulu and Command companies living on meagre rations and minimal water for five months. The conditions, as always, were made the best of but the IED threat made patrolling in Wishtan a grim prospect, testing the nerves and courage of those few men to the limit.

Marine Steven Nethery's gallantry in Wishtan was just one example of the tremendous bravery that was being displayed across the Commando. A patrol from PB Wishtan came under fire as they stopped to uncover what looked to be a command wire for an IED. Caught in the open along a wall that stretched for seventy-five metres, Nethery's section of seven was engaged by heavy and accurate fire from three firing points in the broken ground to the north, with the nearest enemy position containing an RPK machine gun and numerous small-arms weapons at a range of three hundred meters. Nethery quickly bought his GPMG to bear on the enemy, to counter the ambush, but it was immediately clear that the only course of action was to move out of the exposed stretch of ground that comprised the killing area.

One man still lay in the open sixty metres away from where the section consolidated, and he was now the main target of the enemy's fire. Nethery turned to the section commander and simply stated 'I'm going to get him'. He handed his machine gun and ammunition over and urged the rest of the patrol to give suppressive fire. Nethery then ran for sixty metres, in full view of the enemy, towards the most concentrated fire. On reaching the casualty Nethery saw a serious wound to his friend's left leg, and bullets struck the wall and ground around him. Unable to lift the casualty in his equipment and body armour, Nethery began dragging him the full sixty metres back to cover, all the while receiving the brunt of the enemy's fire with no hope of taking cover. The casualty was unable to assist in any way as shock and blood loss began to take their toll.

Nethery informed his commander that vital C-IED equipment still lay unrecovered in the killing area. Wishtan was an area that had seen the greatest prevalence of IEDs in the whole of Afghanistan. Nethery also knew that such a capability could not be allowed to fall into enemy hands and without hesitation he ran back out to recover the abandoned equipment.

During the subsequent withdrawal it was Nethery who firemen-carried his friend the final stretch of two hundred and fifty metres, again under enemy fire, to hand him over to the casevac team.

Once the casualty was clear of the area and on his way to FOB Jackson with the evacuation team, the enemy knew that only six men were now holding the Patrol Base in Wishtan and they set about trying to score a spectacular victory of overrunning a patrol base.

Nethery's patrol commander Corporal Baz O'Connell knew that he had to get back to defend the patrol base, but the enemy were channelling him through an area that had seen huge numbers of IEDs in an area where his patrol were four times more likely to find a device than the next most targeted area in the country. Indeed two-thirds of the command wire devices in Afghanistan were to be found within 500m of the patrol base in Wishtan. Clearing the route with metal detectors as they moved, an intense period of close-quarter battle followed. O'Connell organised his teams with those clearing the route being over-watched by the next man back, who would have to engage the enemy by firing past the man with the metal detector in front because he could not bring his weapon to bear. The closest that the enemy launched an RPG from was forty metres and small-arms duels continued throughout.

A final effort saw the patrol break through the enemy and the sangers and sentry positions were immediately reinforced.

Of the other FOBs, Gibraltar was home to Z Company, who monitored and intercepted enemy forces moving between Sangin and Gereshk. Sited eighteen kilometres north-east of Gereshk along the Helmand River, it was at the apex of a conical-shaped wedge of desert known as the Witch's Hat, which jutted north into the green zone. It was, even by Helmand's standards, an isolated base and the scene of some remarkable gallantry.

Corporal Bradley 'Bugsy' Malone was to be awarded the Conspicuous Gallantry Cross for his 'fighting prowess and gallantry'; Corporal Malone showing this on no fewer than three separate occasions, each one of which would have merited the award by itself.

On New Year's Eve of 2008, and moving towards known enemy positions, Corporal Malone's team had been suddenly involved in an ambush:

'Take cover!'

'Contact! Wait out.'

As he lay in the cover of a shallow ditch, Corporal Malone weighed up the situation. He quickly decided he would lead his section in what they would want to do – attack.

'OK, lads, get your bloody bayonets on! We'll soon sort this lot out.' Then, a moment later, 'Ready?'

'We're with you, Bugsy!'

'Come on then, what are we waiting for...?'

Confronted by the determined and controlled aggression of the Commandos, the enemy promptly fled.

'Bugger that!'

'We'll get them next time!'

On the next time, Corporal Malone showed 'bravery and initiative far beyond his rank or experience' when he displayed a similar level of personal courage and military acumen. While he was firing a 'baby' (51mm) mortar at the enemy, his troop sergeant became isolated and was now being pinned down by enemy fire, unable to move. Once more Malone, accurately gauging the seriousness of the situation, left the safety of a ditch and, braving fierce fire, reached the troop sergeant. Able now to work as a fire and manoeuvre team, the two of them were able to fight their way back to the troop.

The final time was when 10 Troop was moving north of FOB Gibraltar along a narrow but deep waterway. They knew the Taliban were in the area: their task was to seek them out, draw them out, and then take them on. Shortly after midday, the patrol was caught in an immense explosion. The leading section commander was killed instantly, while in front of him, the point marine was blown across a ditch into an open field, very seriously wounded. The third man was Bugsy Malone who, still clutching his GPMG, was hurled backwards off his feet. Shaken and bruised but otherwise OK, he collected his wits, shouted for two men to come forward to help and then grabbed a radio set to call for support.

The Taliban had other ideas, however, and before the dust settled the whole patrol came under fire from four well-prepared firing points: the closest was 100 metres to their right, with another two beyond that and the fourth 200 metres to the left. It was a classic IED-initiated ambush.

'Contact. Wait out!'

Malone, while organising the emergency treatment of the wounded as well as the evacuation of his dead section commander, managed to call for instant air support.

1241:   Contact explosion.
1241:   IED explosion FOB Gib. 1 x KIA and 1 x T1
        casualty.
1250:   Regional Command (South) recommend MERT
        for casevac.

1257:   ETA casualty at Gib – 30 minutes.

1259:   UPDATE Friendly forces came under IED attack, causing 2 x casualties, followed by small-arms fire attack while conducting treatment. FF now treating casualty and returning small-arms fire.

1302:   ETA of casualty to GIB – 5 minutes.

1307:   Casualty now at GIB. Remainder of friendly forces still withdrawing.

1309:   MERT wheels up at Bastion

Continuing to fight the enemy across an arc of nearly 180 degrees, but now with support from other sections moving up their flanks, no other casualties were taken. The object now was not so much to smother the enemy; rather to keep his head down to allow the casualty evacuation to take place.

This desperate time was captured in Malone's eventual citation: 'Undaunted, he immediately took control of the situation and by firing more than 900 rounds, he managed to keep the insurgents at bay while providing covering fire for the evacuation of the dead man.' He was also calling in fire missions from mortars, fixed-wing ground attack aircraft and 29 Commando Regiment's 105mm guns.

Thirty or so minutes after the initial IED contact, a Harrier GR9 homed in with a Paveway IV laser-guided bomb:

1312:   Close Air Support (GR9 Harrier) dropped 1 x Paveway IV on EF FP.

The enemy went quiet. Those caught were dead; those still alive wishing to remain so. The patrol could continue getting its dead and wounded back to FOB Gibraltar.

1315:   Friendly forces out of contact and en route to Gib.

1337:   MERT wheels down at Gib.

1339:   MERT wheels up at Gib. Escorting Apache remaining in support of troops in contact.

However, Corporal Malone and his fellow marines were not out of danger yet:

1355:   Apache has observed suspected enemy setting up a potential ambush as friendly forces are withdrawing. Apache remains overhead, observing.

1422:   All patrol returned to FOB.

Y Company's story was a typical one for the units of Battle Group North. Deployed to FOB Inkerman – two and a half kilometres north along the river from Sangin – in mid-October 2008, Major Rich Parvin's men took over from a Parachute Regiment company consisting of three platoons and a fire support group. The usual attachments that keep every FOB operating were also in abundance during the Paras' time: artillery, engineers, vehicle mechanics, chefs and medics. Additionally, there was also an electronic warfare team and an unmanned aerial vehicle (UAV) detachment that had given them intelligence across the battlefield.

The situation was radically different for Y Company. Like Z Company in FOB Gibraltar, every single commando-trained rank in the FOB, were they a vehicle mechanic, gunner, driver, clerk or store man, would join the company for patrols unless conducting essential tasks within the FOB. The attached ranks from the Commando Logistics Regiment would also prove worth their weight in gold: operating right alongside the rifle sections were fighting chefs; drivers in the thick of the battles; and even the company clerk at the heart of casualty evacuations, by the company sergeant major's side.

In Kajaki V Company had a different challenge yet again. Tasked with protecting the strategically important Kajaki dam with a relatively small force, Major Nigel Somerville and his company seemed to thrive on the relative isolation from the rest of the Battle Group. During their tour they too had to deal with an ever-present and deadly IED threat and an active enemy who were keen to engage the company with small-arms fire and indirect weapons whenever the opportunity arose. Throughout they performed with considerable bravery and tenacity and kept the insurgents well away from the site of the dam.

In mid-December 2008 Victor Company conducted a patrol to the northern FLET typical of the format of operations in Kajaki. As ever it was a company-level patrol including the ANA platoon which meant that anyone in the FOB that could deploy on the ground did so, leaving a minimal security force in the FOB. Once again the fact that every man was Commando-trained first was underlined, with recce operator, engineers, chefs or vehicle mechanics all required to do their share of patrolling. The patrol was intended to push further north than normal to disrupt a suspected Taliban centre of mass. The company's fire support group deployed early to clear their positions on the numerous high features that dominated the northern area. They were followed by Patrols Troop and the ANA with their mentoring team.

The efforts to beat the Taliban's very capable early-warning screen by leaving before first light worked and Patrols Troop, under the command of Captain Tristan Finn RM, managed to establish themselves on Essex ridge before the enemy spotted them. Once they were spotted, however, the reaction was ferocious; it was clear the Taliban were extremely concerned with the patrol's proximity to their lines. A significant battle ensued with the company at one point being engaged from a 270-degree arc by a combination of DShK heavy machine gun, rocket-propelled grenade (RPG) and small-arms fire testing the company and its commander to their limit as they coordinated mortar fire from the FOB, GMLRS from further south in the Sangin Valley and close air support.

Through the use of indirect firepower the company were able to defeat the enemy, inflicting significant casualties and sending the message that they possessed a greater freedom of manoeuvre than they were prepared for. Fortunately despite a number of exceptionally close calls, including MA Marine Steve Husbands being struck by a 7.62mm round in his body armour and the FSG's Jackal vehicles bearing numerous scars of how close the fire had come, the company returned to the FOB without suffering any casualties.

With Battle Group North's mission in mind, ISTAR had identified a number of Taliban havens nestled in the Sapwan Kala district, twenty or so miles north-east from Sangin, further up the Helmand Valley. This area, on the east side of the Helmand River, contained many compounds housing both opium-processing plants and IED factories.

These were not cottage industries but full-scale production lines. It was fully recognised that if these two linked industries could be disrupted or, better still, destroyed, then less money would find its way into the Taliban coffers, while fewer IEDs would be available to maim and kill both civilians and the security forces.

Sapwan Kala was on the river and its associated shores and flood plain, which were just over one kilometre wide. This had allowed the enemy occupants of Sapwan Kala to consider, with some justification, that they were exposed only on their eastern flank. How wrong they were.

Operation Diesel, phase 2A of the *Aabi Toorah* series of operations across the province, was one of the largest discrete operations in 45 Commando's history. It was to be an ambitious helicopter raid supported by ground manoeuvre elements including Royal Marine forces, the Brigade Reconnaissance Force, and Afghan security forces.

The operation was to strike deep into an enemy stronghold in the Upper Sangin Valley, with a rapid build-up of combat power into the target areas that created an overmatch that would end in all identified enemy command nodes, logistics, IED facilities and narcotics targets being destroyed or disrupted. Opium and refining chemicals capable of producing the equivalent of £50,000,000 worth of heroin were also seized, an action that was to have an enduring psychological and physical effect on the enemy.

Preliminary plans for the operation, begun early in January 2009, indicated that the helicopters likely to be available for the lift would be Chinooks, Sea Kings and USAF CH53 Sea Stallions. This would mean a simultaneous lift of 190 personnel; Lynx would also be available for carrying the Tac HQ. Three waves would be enough to place the whole force of 537 on the ground.

Lieutenant Colonel Morris planned that *Operation Diesel* was to be conducted in eight distinct phases, beginning with phase one: the rebalancing of his forces in order to position them for battle preparation, and the undertaking of specific-to-task training.

For tactical and logistical reasons, Morris could not employ his whole Commando, as he remained committed to the enduring efforts of their various locations. It was, therefore, necessary to borrow L Company from 42 Commando to work alongside his own Y and X Companies, while a number of his own men were in turn redistributed among forward operating bases to replace men borrowed from them.

Phase two of the operation was battle preparation and shaping operations, with phases three, four and five the helicopter lifts into landing sites Willow, Cherry and Oak. Prior to the landings, the Brigade Reconnaissance Force – commanded by Major Chris Haw – would secure two of the three landing sites, Oak and Cherry, and provide over-watch; while 1 PWRR's armoured vehicles would secure the offset landing site, Willow, which would be used to build up the assault force.

The Brigade Reconnaissance Force (BRF) was the latest unit in a long line of reconnaissance, mountaineer and cold weather experts, and offered the Command Support Group – its parent unit – the eyes and ears the Commando Brigade needed in extreme conditions. Just one such function included the Radio Recce Teams who, among many other duties, monitored the Taliban's communications. Also attached were the explosive ordnance disposal men from 24 Regiment Royal Engineers, and 148 Battery Royal Artillery, who would normally control naval gunfire support. For *Operation Diesel*, the BRF was 106 men strong, with six snipers among that number.

Based on a number of targets that had been – and were continually being – identified for *Operation Diesel*, the CO's plans developed, with Jim Morris's proposed force growing to over 700 troops. Drawn from Y and X Companies of his own battle group, L Company and the Brigade Reconnaissance Force, they were also augmented by Afghan Security Forces, armoured infantry and close reconnaissance teams from 1 PWRR, plus Battle Group Centre South's Armoured Support Group.

*

As the planning became more detailed and complex, it was clear that advance force operations deliberately aimed at confusing and deceiving the enemy would need to be conducted. These would then be followed by the carefully coordinated securing of landing sites, before the helicopter-borne troops inserted prior to a rapid assault on the multiple targets.

To carry out the advance force operations, to secure two of the three landing sites and to provide over-watch and fire support from high ground to the east of the three target areas was the task allocated to the Brigade Reconnaissance Force.

The Brigade Reconnaissance Force had been given an excellent handover by their equivalent in the Parachute Regiment – the Pathfinders – since when they had expanded their remit to include working with an Afghan task force of local men, trained as scouts and mentored by the British. These men had been drawn from the Afghan Territorial Force and were considered to be by far the best-trained of all local troops. In addition to close reconnaissance, They were deployed on such esoteric duties as human terrain mapping and tribal dynamics for they were more familiar with the locality, and could talk more easily with locals, enabling them better to gather this sort of information. Haw was an admirer.

On 5 February, at 8 p.m., Lieutenant Colonel Morris held his orders group and rehearsal of concept at Camp Bastion. The objectives to be cleared were along the river's east bank and were labelled South, West and East. Each consisted of a number of compounds and farm buildings, and were situated two or so kilometres away from an area known as Brown Rock. They were flanked to the west by the river and to the east by the mountains.

Frustratingly, there was then a weather-induced twenty-four-hour delay. The weather was settled in the vicinity of Camp Bastion, but further down Helmand River's valley it was decidedly uncertain. Morris did not want to risk warning the enemy of any impending action by getting the helicopters there, only to have them turn around, unable to land due to unsuitable weather conditions.

Finally, on 6 February, the Brigade Reconnaissance Force left Bastion in Jackals, accompanied by Afghan Forces with their British mentors. Major Haw's detailed instructions were simple, but were to prove far more complicated to execute.

The unit was to deploy via ground manoeuvre, maintaining a discreet presence and minimising the extent to which dickers would be able to guess their route. They were to secure landing site Oak, establish over-watch of Cherry and seize and hold in over-watch the key terrain of Brown Rock, known to the Afghans as Mali Ghar, which overlooked, from 1,223 metres above sea level, the Sapwan Kala area. They were also to site mortar positions then, during the contact, provide indirect fire control while preventing any enemy leaving or reinforcing from the north. It was a typical reconnaissance task for a typical commando-style raid.

Because part of the plan was to deceive the enemy as to their final objective's location, it was in fact necessary for the Brigade Reconnaisance Force to be seen in the early stages of their journey. As Afghanistan is known for its eyes and ears – it is generally believed that typical friendly forces can expect to be covert for about one night – this was expected to be easily achieved.

To begin with, Haw led his team from Camp Bastion to Gereshk, from where they headed into the desert and hills as far east as Kandahar. He was, at this stage, following the route that had been used to get the turbine to the Kajaki dam in September 2008. The area was known to harbour Taliban in large numbers and, sure enough, they were to take an interest. Now, by taking the long easterly sweep into the foothills before turning back north-west towards the Upper Sangin Valley, via the notorious bottleneck mountain pass at Ghowrak which lead to the wide sand and gravel plains of eastern Helmand, Haw was certainly advertising his presence.

Initially, the squadron got away unscathed but, as the BRF and its accompanying units moved into their final lying-up position before the objective, with a flash of light, acrid smoke and spray of sand and stones, a 107mm rocket suddenly landed close by.

> 1739:   BRF returned fire on the enemy forces position with heavy machine guns.

Job done, the BRF still spent an anxious night knowing that the next morning they were due to transit the narrow Ghowrak pass and from there north – unseen they hoped – to their target area.

The next day, all went well until the approach, past compounds, to the gorge's narrow defile. Suddenly, about 500 metres short of it, the patrol was temporarily stunned by a huge explosion. When the dust and smoke had cleared they could see that a Jackal was lying damaged, two of its crew seriously wounded. Men rushed to their aid, even before the echoes round the gorge had faded. The time was 0951: less than twenty minutes later, the standard NATO nine-liner medevac request had been transmitted to Bastion, describing the state of the two men: one urgent and one priority.

To add to the general air of stimulation, the enemy now engaged with mortars and phosphorus. Swiftly assessing the baseplate's position, fire was returned while a friendly B-1B bomber, far overhead, broke many of the rules and dived to 5,000 feet in a show of force. The huge aircraft was so impressively big that even at that height it blocked out the sun for the troops on the ground.

*

Apaches, too, were soon orbiting and observing and, if nothing else, were keeping the Taliban snipers at bay while the two casualties were stabilised for evacuation by Chinook. On arrival at Bastion's hospital, the consultant surgeon was said to express concern at the facial disfigurement suffered by Sergeant 'Banjo' Haigh. 'Don't worry, doc,' had come a reply, 'Banjo always looks like that.'

The other casualty, Kenny Brewster, was in good spirits but it would take rather longer for his injuries to heal.

After the successful medevac, it was time to deal with the Jackal, and aim yet more devastating fire at the identified enemy compound and base-plate position.

1148: SITREP One guided bomb dropped on to compound, main compound intact, wall damaged. Enemy forces laying IEDs in front of BRF. B-1B bomber and Apache remain in support.

1417: Friendly forces continue to Barma the Ghowrak Pass and on completion the intent is to deny the Jackal. All equipment has been removed.

1456: Jackal has been denied by Apache using 30mm and Hellfire.

The Brigade Reconnaissance Force could now continue through the pass. For one kilometre they moved slowly along the gorge, the hair on the back of everyone's neck bristling. All knew that their progress was being closely monitored. They needed to move fast to get out of the area as their slow speed gave the enemy every advantage, but they had to move slowly as the Barma team, out in front – on foot and exposed – searched diligently and painstakingly for further hidden ordnance. It was daylight and nerve-shattering as the brave team inched them slowly forwards.

As they moved away from the narrow defile, the enemy communications that had gone quiet during the engagement were now suddenly more vocal:

'They are going to attack Sangin.'

'No, they go to Kajaki. Same route as the turbine.'

'You're wrong. They are heading for Malmand. They will attack there.'

The Y Squadron Radio Recce Team longed to assure them that good as they were, even the Brigade Reconnaisance Force could not do all that in a single night.

The deception appeared to be working as the Brigade Reconnaisance Force feinted westwards long enough to confuse any dicker before turning towards the steep slopes of their last objective. Finally, under the cover of darkness, the BRF and Afghan forces moved up out of the wadi to gain the slightly higher ground beneath Brown Rock, their final objective.

It had been a gruelling journey, but one eased considerably by the procedures that had got them there. Long before setting out from Bastion, the whole route had been recced by map and helicopter by mountain leader Sergeant 'H' Howley. His time spent on reconnaissance had not been wasted, as the route selection received the thumbs-up all round. When it came to the execution, he spent the whole operation with the grim task of being in the lead vehicle.

Huddling down to escape the cold, while the light of a full moon and a strong wind that hurried the occasional cloud over the moon's face caused confusing shadows to flit across the scrub, Major Haw radioed his successful insertion into the area to Battle Group North's HQ via satellite comms.

The HQ staff could stop holding their collective breaths. The BRF's circuitous convoy from Bastion had not been without incident and could so easily have been compromised – and crucially, delayed – at almost any point along the journey, but it was now safely in place, and on time.

Next on the BRF's agenda was to recce the base-plate position for 45 Commando's mortars. Captain Charlie Breach – another mountain leader – now led his heavily laden team, at night, 800 feet up the near-precipitous southern slopes of Brown Rock from valley level. Each man carried well over 100 pounds in weight, made up of personal equipment, Javelin missiles, sniper rifles, forward air control radios and all the equipment needed for carrying out a lengthy over-watch.

Shortly after setting off for the summit, the feeling among Breach's 1 Troop that all was going too well was suddenly emphasised when shots rang out. For those back in the valley, however, and despite the sudden halt in proceedings, no 'incomers' were reported, no 'Contact! Wait out!' came over the radios. Silence.

Slowly, the details were received in the unit's HQ: an Afghan interpreter had gone down with heat exhaustion and, in accordance with Afghan standard operating procedures, to attract attention and to avoid being left behind, had fired his AK-47 into the air, completely breaking the troop's cover. Surprisingly, and luckily, these lone shots in the darkness did not alert the enemy and 1 Troop was able to continue their climb.

By midnight – two hours before the first helicopters were due – the Brigade Reconnaissance Force's tasks were complete.

All was now in place for Operation Diesel's third phase. L Hour – the time that the first helicopters are scheduled to land – was 0200. Before that, however, X Company, made up of of 115 marines and some Afghan soldiers, was to be pre-positioned at landing site Willow at midnight – or L minus 120 – to enable a more rapid build-up of combat power close to the three objectives. They would then wait at Willow while L Company was lifted straight from Bastion to landing site Cherry, and the Commando's Tac HQ and a mortar team were lifted to Oak, both teams touching down at 0200. It would then be X Company's turn to be lifted again, this time into landing site Oak, to join the Tac HQ and mortar troop, at 0220. The key idea was that when X Company became the second assault wave into Oak, there would be a minimum delay between the two of just twenty minutes. The third wave would be Y Company – with the Regimental Aid Post – assaulting into Oak after X Company had moved out. There, close to the expected battle, they would remain in reserve until called forward. A mobile air operations team that had accompanied the Brigade Reconnaissance Force were ready to bring in the aircraft to Oak.

Landing site Oak lay in a bowl beneath the southern slopes of Brown Rock, now under the protection of Captain Breach's fire support troop. With their heavy weapons and night-vision equipment, they had been carefully positioned so that they could also keep an equally close eye on the other landing site, Cherry, which lay three kilometres to the west on a flat bulge of the eastern edge of the river's flood plain. All was in place for the assault troops' arrival.

The Brigade Reconnaissance Force's deception plan had not been the only one in operation as, during the preceding week, men from 1 PWRR and the Armoured Support Group in their Vikings had deliberately faked an operation into an area to the north of the targets. This, too, was intended to divert the Taliban's attention away from the intended area of assault, thus strengthening the desired element of shock and surprise.

Having driven across the desert to the west of the objectives, 1 PWRR's armoured vehicles had secured landing site Willow, which was around eight kilometres north-west of the objectives. Once the western flank was secure, their duties then were to screen the target area to prevent enemy reinforcements arriving, while intercepting any Taliban attempting to flee. The Vikings of the Armoured Support Group had moved into the south, ready to provide resupply and casualty extraction, if required, while others of the Armoured Support Group, and elements of the PWRR in their Scimitar tracked reconnaissance vehicles had moved into positions to the south and west as deception and blocking forces.

At Camp Bastion, well before that first pre-positioning flight, three RAF Chinooks, two Royal Navy Sea Kings and two Sea Stallions from the USMC's Special Marine Air Group Task Force sat on the airfield apron, their lights red and dimmed, while the crews conducted their pre-flight checks. Close by, 537 men of Task Force Helmand were going through their own last-minute checks, overseen by their individual NCOs and troop commanders. There was the usual sense of expectancy and trepidation among those about to go into battle, now heightened by the cool, sand-filled wind and the moon's shadows that moved silently across the surrounding desert.

The assault was due in three waves, with 42 Commando's L Company leading into landing site Cherry, which was on the river bank two kilometres south of the West objective. Before that, though, X Company's lift to Willow had to take place.

A few minutes before midnight, the 115 men of X Company with more Afghan colleagues, their own Tac HQ, more mortar teams, a combat camera team (to record events for historical accuracy and should evidence ever be needed) and two quad bikes embarked on to the waiting helicopters. Those in the Chinooks filed up the rear ramp in the reverse order to which they would disembark. With the helicopters already 'burning and turning', the heady and evocative smell of burnt kerosene, the whine of the twin engines and the dust-laden air would remain in memories of the battle for years to come.

The marines lucky enough to be on first tried to take their places on the webbing seats either side of the aircraft's fuselage: bulked out with weapons, water and ammunition it was well nigh impossible to sit back in the canvas bucket seats, while those who had no seat stood awkwardly. A cool draught ran the length of the cabin, bringing with it the taste of Afghan dust. Nobody could talk above the noise of the idling engines. It didn't matter though, as it was a short-haul flight. As the helis rose, swinging round to orientate themselves in the direction of travel, those that could, twisted to look out of the windows, watching the moonlit desert flash past, not many feet below. The pilots, looking out through their night-vision goggles, muttered that it was almost too bright, except when the moon's glare was masked by cloud.

All too soon, the marines could feel the aircraft begin to slow, and then flare out to drop forward speed even more. Through the open stern ramp the dust billowed in.

Eyes shut. Breath through the nose. Welcome to landing site Willow.

'Out! Move! Move!' The loadmaster's shout and arm-waving were welcome. It was a relief to get down the ramp and away from the 'bullet magnet', and back to the medium the marines understood. Down into a fire position. All-round defence. Practised and practised. Another wave of orifice-clogging dust as the helicopters took off again, and then silence. X Company had some minutes before they would become the second assault wave, this time into landing site Oak. There, they knew, they would be met by the mobile air operations team who would be waiting to lead them into their assembly areas and start lines, and the start of the assault.

*

Meanwhile, back at Bastion, the aircraft touched down in the same spots they had left minutes before. It was time for the first assault wave to embark in order to make their 0200 L Hour at landing sites Cherry and Oak. L Company's 175 men plus Tac HQ, the mortar sections and sections of the Afghan forces were led across the darkened tarmac, forty towards each Chinook, ten for each of the Sea Kings and twenty-five for each Sea Stallion: an initial wave of 200 personnel. One Sea Stallion and two Lynx would divert to LS Oak with the Commando's Tac HQ and a section of mortars.

With the unmistakable beat of the Chinook's twin rotors masking the whine of the gas turbines, the nine aircraft of the first assault wave streamed into the air to disappear north-east towards the moonlit landing site Cherry.

They were on time but they were not on target. The lead aircraft, followed by the others, touched down 500 metres short of their correct destination. This was not an auspicious start to such a complicated insertion. Cries of, 'Where the fuck are we?' rang out across the huddles of men as the now-empty helicopters lifted off into the moonlight.

'Typical Crabs.'

'Women drivers more like.' Which was true, as the lead pilot had been a woman who was, reportedly, mortified.

'We're not where we are supposed to be!'

An officer took charge and swiftly compared the ground to his map. 'Its only five hundred metres, lads. Not even a yomp!'

The flight of mixed helicopters did not return to Bastion immediately. Having dropped their passengers at Cherry, the aircraft then flew swiftly across the nine kilometres to collect X Company, plus their attached Afghan colleagues, from Willow. They were now the second assault wave and this time landed at Oak a mere twenty minutes later, where the Commando's Tac HQ was already waiting beneath Brown Rock. Their target was the South objective, two kilometres to the north-west of the landing site.

Meanwhile, L Company had moved off Cherry, along the east bank of the Helmand River to begin the systematic clearing of their West objective, while also securing the vital crossing where the river and its flood banks narrowed. The enemy was now denied the freedom to move in reinforcements or to escape.

X Company was by now fast approaching the South objective and had engaged, fighting through and searching the enemy positions.

Finally, Y Company's 110 men landed at Oak, held in reserve should the others need help, and to be prepared to invest force into the East objective area when it was needed. Within striking distance of their own objective, Y Company's company commander could hear echoing round the mountainside sporadic shots and exchanges of small-arms fire throughout the morning, but not the heavy defence they had been prepared for. It was clear that several groups of Taliban had moved out to meet phantom threats as they had hoped, and that they were only relatively lightly armed.

Four kilometres to the north-west, L Company with their Afghan colleagues were forcing their way through compounds, searching for the enemy who, totally caught by surprise, were choosing to drop their weapons and flee, leaving behind all manner of paraphernalia associated with the manufacture of narcotics.

Exploding their way into a compound believed to contain narcotics, marines surged through the walls, weapons raised to their shoulders, night-sights switched on, prepared to shoot any armed insurgent first and then ask the questions. The first of two processing plants was discovered in a chain of compounds with all the vital ingredients for the drugs trade having been abandoned as the enemy fled. There were large vats for refining opium; presses and chemicals used in the production of heroin were in use; while bags waiting to be filled with the drug were stashed along the walls. Sixty kilograms of wet opium were waiting to be processed in the one compound alone.

The swift progress did not go entirely unchallenged by the resident Taliban, however:

0705:   Friendly forces (fighting through West objective) found two enemy wearing assault vests and carrying AK-47s and a belt-fed weapon. Engaged by GPMG and sniper.

0749:   Four enemy positively identified with weapons moving along a wall between compounds. Engaged. Battle damage estimates four enemy KIA.

0823:   Lima Company received a burst of small-arms fire. Engaged by sniper. One enemy KIA.

0853:   Two enemy identified engaged with small-arms fire. Two enemy KIA.

0858:   Enemy small-arms fire. Returned.

X Company was not having it all its own way, either, as pockets of resistance began to spring up. Part of the Taliban's defence mechanism was, as was so often the case, purposefully chosen warrens of compounds, ditches, narrow alleyways and tunnels. Random counter-attacks were launched against X Company from these and the Afghan forces returned fire, supported by Apache attack helicopters accurately targeting the enemy firing points. The attacks were soon quashed and the area secured, with the enemy again abandoning compounds and fleeing. And so the day continued, with incoming enemy small-arms fire being met with small-arms fire, Javelin and mortars. If the marines of Battle Group North had hoped for close-quarter combat they were disappointed. The enemy, shocked into action and caught completely unawares with no coordinated plan for mutual defence, had instantly decided to make a run for it, only engaging their assailants at a distance, and almost always to their own disadvantage.

Having been held in reserve for the majority of the morning and listening to their fellow troops engaged to the north, Y Company was at last cleared to launch their attack on East, the final objective. With no reserves now needed, they moved up through the area previously secured by X Company and advanced north to the collection of compounds which comprised their target. Manning mortars, heavy machine guns and with a sniper team, they also moved on to high ground beneath Brown Rock to the east to provide another layer of over-watch.

They were 200 metres away from the first compounds and had just climbed on to the high ground when an RPG buzzed over their heads, fired by a man who had come out of an alleyway. Fortunately the grenade was a blind and failed to explode. The marines opened up on him in response, but as another firing position with a PKM heavy machine gun was also firing at them, they guided in an Apache helicopter that attacked with its 30mm cannon.

In one compound they found ten or fifteen barrels of wet opium cooking, which was the most the marines on that mission had ever seen. Once they had finished searching, it was all put into one compound and destroyed.

The snipers concealed among Brown Rock's lower slopes and gullies, as well as those lying flat on compound roofs, were kept busy spotting the enemy as they fled from the area. In truth, 'Terry' had little option, realising that by the strength of numbers and equipment ranged against him it was clear that death or retreat were the only two options open. Because of the distance involved, the mortar teams and snipers were having a field day, and the general feeling was one of wanting to send the Apaches away as the lads were having such a 'hoofing' time of it.

As the battle for the three targets fizzled out and with the insurance of a B-1B bomber permanently overhead and an Apache on call, equipment – so vital, financially as well as kinetically, to the insurgents – was dragged out into the open or tabulated where it stood. The finale was the uncovering of the largest and most complete processing laboratory to be found up to that date.

After the day, described as one of minor skirmishes but huge finds, all three companies yomped south-east out of the green zone, just before last light. The extraction route, still guarded by the Brigade Reconnaissance Force from the peak of Brown Rock, led over the hill's southern slopes and into landing site Oak. Within forty-eight reasonably trouble-free hours, everyone was back in Bastion enjoying a well-earned … soft drink.

In line with military procedures, the BRF's return to Camp Bastion had not been along the same route as their outward journey. Nevertheless, the return journey was still regarded as a treacherous one, initially in company with the Vikings as far as an obstacle crossing east of Sangin, which had been secured by US forces. The squadron then hand-railed the terrain features south along the Helmand and back to the bridges across the river and canal immediately to the east of Gereshk. From there it was westwards down the A1 road until the turning south for Camp Bastion. Although there had been no enemy interference, anxiety had been permanently present.

With every possible good reason, Lieutenant Colonel Jim Morris was extremely happy. *Operation Diesel* was a bold and hugely successful operation that had demonstrated quite clearly that ISAF and Afghan forces could strike when and where they chose, with speed and ruthless precision, at the very heart of Taliban bases. The resulting disruption of the enemy and his infrastructure, and the impact on the closely related narcotics activity that was achieved in the action would contribute directly to the gradually improving security situation in the Upper Sangin Valley.

Throughout, the enemy had been caught on the hop, and remained on the back foot for the entire operation. With no casualties sustained on the day, the marines knew it was a job well done. In one night, they had reinforced Government control over the Upper Sangin Valley; denied the enemy safe havens and freedom of movement; removed a considerable number of weapons and ammunition; discovered a motorbike that had been modified for a suicide attack; disrupted and removed enemy forces and the drug chemists; destroyed large quantities of narcotics, including 1,295 kilograms of wet opium which, as heroin, would have had an estimated street value of over £6 million; destroyed four significant drug factories and laboratory equipment, including four large vats and drug presses; and destroyed 5,000 kilograms of ammonium chloride, 1,025 litres of anhydride, 1,000 kilograms of sodium chloride and 300 kilograms of calcium hydroxide. Oddly enough, it was these chemicals that were the real find. Opium was relatively easy to come by, compared to the large amounts of chemicals needed to process the drug. These latter quantities would have been sufficient to produce heroin with an estimated street value of around £50 million.

Morris's men had exposed the direct link between the Taliban and the narcotics trade. The Brigade Commander himself summed up the success of the operation, stating that *Operation Diesel* had been a clinical precision strike, supported by strong intelligence, which had had a powerful disruptive effect on known insurgent and

Above: Kate Nesbitt shortly after saving Jon List's life.

Below: John List's Bergen.

Left: The Fairbairn-Sykes Fighting Knife carried by the Commandos. Royal Marines, who know it colloquially as a Commando Dagger, are trained in its use.

Below: Morning sick parade.

Above: Lieutenant Colonel Charlie Stickland's Orders Group at Camp Bastion for *Operation Aabi Toorah 2B*.

Below: 42 Commando's planning team during *Operation Array*.

Above: Jackals at dusk.

Below: Javelin missile – 'on target'.

Left: A member
of a 'Barma'
team in action.

Above: 'Danger close' air support.

Below: Action following the BRF minestrike in the Ghowrak Pass.

Above: 42 Commando Christmas cheer on completion of *Operation Sond Chara*.

Below: 42 Commando Christmas dinner. Lieutenant Colonel Charlie Stickland on the right.

Above: 'Dog tired' during *Operation Sond Chara*.

Below: 'Hearts and minds' at work.

Above: Marines from 42 Commando during *Operation Sond Chara* enjoying the last of the sun before the 'Somme-like' conditions took over.

Below: Decompression. Two days of enforced banana-boat riding and beer drinking in Cyprus on the way home.

narcotics networks in the area. The success of the operation was seen as a significant boost to the Afghan authorities in their fight against the drugs trade, and he vowed that the combined ISAF/Afghan forces would continue to take every opportunity to strike at the connection between the narcotics trade and the Taliban, the product of which brought misery to the Afghan people.

Throughout its tour, the men of 45 Commando, forming Battle Group North, faced a determined insurgency and, while enemy forces attempted to dominate and influence the lives of local nationals, much of their activity was suppressed. It also allowed progress to be made in training the Afghan National Army and police, while also disrupting the narcotics trade in Helmand. The battle group also conducted a series of intelligence-led, targeted operations against the Taliban by surging troops into the target areas, each ending in the discovery of weapons, ammunition and explosives. These efforts often resulted in direct and lethal engagements but overall created a stability that allowed schools, clinics and new shops to open: all important indications of life returning to some degree of normality for ordinary Afghans.

CHAPTER SIX

KANDAHAR AND THE FISHHOOK

Then, a sudden flash of light, an eardrum-blasting crack, a cloud of black and white dust and the horrific 'red mist' as eight pints of blood turned to spray. Everyone instinctively ducked although it was too late to avoid some unidentifiable bits of Taliban that came spinning across the gravel. As the smoke cleared and marines stood to shake their heads clean of dust, everyone scanned the ground around them – mercifully, there were no British casualties, no British body parts.

For the duration of *Operation Herrick 9*, 42 Commando was detached from the 3rd Commando Brigade to support instead Regional Battle Group South (RBG(S)), working directly with the Dutch-led Regional Command (South) HQ (RC (S)). They were on rotational command as a NATO post across all six provinces under Regional Command (South)'s control: Dāykondi, Orūzgān, Zābul, Kandahar, Helmand and Nïmrūz. As it turned out, however, Stickland's 'Smiley Boys' of 42 were under the operational control of Task Force Helmand rather often.

Capped for political reasons at 436 marines, despite its formed strength of over 800, 42 Commando manoeuvred across most of southern Afghanistan executing operations that ranged from independent, deliberate, night-time helicopter-borne strikes to set-piece attacks in conjunction with other units of the Commando Brigade, or Canadian and Dutch forces. These included the clearing of Taliban strongholds, assisting with security for voter registration and the delivery of humanitarian relief before winter came to Kandahar, Orūzgān and the southern extremes of Helmand.

The commandos of 42 were often the first foreigners that many local nationals had met and the importance of leaving behind an enduring good and positive image was impressed upon the marines by their commanding officer.

Despite being capped and going on to lose one company to Task Force Helmand, it was still vital for 42 to operate with three manoeuvre units. In consequence, Stickland reorganised his Command Company, containing the anti-tank, grenade machine gun, heavy machine gun and recce troops, and formed them into the Commando Reconnaissance Force (CRF) under Major Adam Crawford. As he always needed three companies for helicopter assault, Stickland also gave Crawford plenty of Afghans to make up the CRF to 110 strong, divided into three troops and a Fire Support Troop, mounted in over twenty Jackals.

42 Commando had declared itself operational on 22 September 2008 and wasted no time in proving this status. On that day, under the command of the Canadians working with the Kandahar Provincial Reconstruction Team, L Company – under the command of Major Rich Cantrill – proudly laid claim to being the first into business on *Herrick 9*.

Deployed into the rural fringes, south-west of Kandahar city with a small band of Afghan National Army warriors, L Company investigated an area not regularly patrolled by coalition forces. The marines conducted mobile, vehicle-borne operations – for which they had had no preparation during pre-deployment training as they had not been expected to operate in this manner – to carry out their mission, that of disrupting insurgent activity during Ramadan and the festival of Eid, and to increase 42 Commando's situational awareness of this key area for the benefit of future Canadian operations.

Having 207 men in up to seventy-two vehicles was a bit of a culture shock for marines more used to patrolling on foot – never before had they deployed on commando operations with equipment such as the huge DROPS vehicles, and a light-wheel tractor.

Clearing their way through insurgent territory in vehicles that, because of their limited mobility, were restricted to the main routes was testing, yet this first operation offered a useful glimpse towards the future, and lessons were learned. It was also a tense time during which four of the Mortar Troop were injured in an IED strike on a Vector 'protected' patrol vehicle.

After two weeks on the ground, and little direct engagement with insurgents, they prepared to link up with the rest of the Commando. Their dynamic patrolling and employment of an intelligent and balanced attitude towards the local nationals helped neuter any enemy influence by offering a positive message of support to the weary population.

By 5 October, Major Rich Cantrill's L Company Commandos and his attached ANA warriors made a difficult move deeper into the 'Bandit Country' of Zari Panwayi to link up with the rest of 42 Commando. This was a tense time during which four of the Mortar Troop were injured in an IED strike on a Vector 'protected' patrol vehicle. *Operation Array* saw Stickland's Commando operate under command of the Canadian Task Force Kandahar and team up with the 3rd Battalion Royal Canadian Regiment (3RCR) who were to become close brothers-in-arms throughout their tour. The operation was initiated by multiple moves with 42 Commando seeking to overload and dislocate their foe in what was known to be a Taliban staging post and command control node centred on the Zalakham village, some fifteen kilometres to the south-west of Kandahar. The purpose of *Operation Array* – the start of which had, coincidentally, coincided with the Taliban attacks on Lashkar Gah in October 2008 – was to disrupt and dislocate insurgent operations in the eastern district of Panjvai in order to prevent their malign influence on Kandahar City.

The attack started with a Recce Troop aviation assault on to a knife-edge high feature and 'owning' the high ground and vectoring supported AH on to targets as they emerged. There then followed a three-pronged approach – L Company from the north, CRF and Commando Tac from the south-west supported by ANA troops ferreting about for information and insurgent activity and, much to the surprise and disconfiture of the Taliban, K Company aviation assaulting directly on to the individual troop objectives.

All helicopter assaults come with a hefty dose of adrenaline. They are associated with a heady cocktail of opposing feelings and emotions that include aggressive power, trepidation and tremendous excitement. The sense of aggression comes from the helicopter's ability to touch down on or near an objective with significant shock to the enemy, while trepidation comes from the knowledge that, although they should not be aware of the approaching aircraft until seconds before they land, the enemy could be expecting the assault and thus have prepared his defences accordingly.

These 'tri-polar' emotions were very much in evidence among the Black Knights of K Company as the Chinooks descended into Zalakhan, a village with a notorious and lengthy history of insurgent activity, stretching back to the Russian occupation. As recently as two months before, US troops had been in contact for four days in and around the village.

Eventually, after the usual last-minute hiatus of changing plans, loading and insertion landing site rethinks, the company finally embarked.

While the aviation insertion achieved both operational and tactical surprise with the Company's arrival at dawn – suddenly and within the Taliban's defensive perimeter of sentries and established positions – and had caught the insurgents absolutely unawares, the insertions did not go according to plan. The Tac HQ and Fire

Support Group were rapidly engaged by the enemy when they found themselves dropped off on their own front line, while 6 Troop had also been dropped off at the wrong grid coordinates, but had quickly moved themselves forwards and into the village. Steadily, the ground elements manoeuvred carefully forwards through an eerily silent village – always a sign that something was amiss – towards a pre-chosen, centrally located compound that would be used as a forward operating base.

As K Company's sergeant major and Fire Support Group commander were trying to locate the patrol base, the insurgents laid down a disconcerting rate of fire – if not accurate then still worryingly close – using Kalashnikov machine guns and RPGs. Realising their predicament and with fierce courage, 6 Troop responded in kind by manoeuvring swiftly into the killing zone through covering fire, pepper-potting to provide depth to the threatened, embryo patrol base.

The insurgents continued to engage throughout the afternoon, but when 5 Troop began counter-attacking from the east, supported by particularly accurate mortar fire, the insurgents, appreciating that they were all but surrounded, began to give ground and extract.

Corporals Tom 'Webby' Webster, Paddy Cochran and Marty Stronach fought their sections with tenacity. One section would go to ground to support another on the flank, then they would swap roles as the sections leapfrogged forwards in turn. A non too steady stream of orders, advice, encouragement and questions were barked, sometimes breathlessly, over the PRR net:

'Webby. Got you covered, mate. My gun group's to your left. Shout when you reach that fucking compound. You'll be in dead ground.'

'Cheers, oppo! Come on, lads. Move it!'

Zigzagging, the marines leapt and dodged over the uncertain ground and trenches of the green zone, their heavy body armour, Camelback water carriers and spare ammunition not helping their agility.

'Paddy. We're on the target. Who's got the fucking ladder? I need a sniper on the roof. Now!'

'Watch out. The bastards are in the ditch 150 metres to your left.'

'Seen. Watch my tracer then give them bloody hell.'

'Roger.'

A twenty-second pause, and then:

'Fucking ace!'

'Good shooting. Lovely. No need for a stretcher.'

'Shovel any good?'

Fighting through, and over, the compounds, the marines completely overwhelmed the insurgent threat, while overhead, support came via a variety of airborne surveillance systems as well as from the ubiquitous Apache attack helicopters.

It took five days, but at the end the ISAF troops, who had been in contact with the enemy throughout, had acquired a huge amount of intelligence and uncovered a mass of bomb-making equipment. By dislocating the enemy's early-warning screen, investing the area simultaneously from a variety of directions and by forcing the Taliban into the open, numerous compounds were identified as targets for subsequent precision strikes by Canadian forces.

Following these engagements, the battle damage assessments indicated that while the two commando companies had, in kinetic terms, defeated the Taliban in Zalakhan, this was nothing compared to the longer-term disruption to the enemy achieved by the exploitation of the village. In almost every compound, and during subsequent discussions with the locals – who began to reappear in the outlying fields following the first day's fighting – evidence of insurgent activity was rife. *Operation Array* revealed some of the largest finds of insurgent paraphernalia and equipment within Kandahar Province in recent years: eleven constructed IEDs, many bomb-making components, RPGs, fuses, detonation cord, grenades, weapons, ammunition of many calibres, optics, communications equipment and basic first-aid kits. Anything that was found was destroyed.

Incriminating documentary evidence was also recovered from two main locations within the village and, to a lesser extent, from many others. What was particularly heartening was that very little of all this *materiel* was hidden, such had been the insurgents' confidence that they were safe in their strongholds, and compounded by their haste to escape. Attempts to delay, disrupt or even prevent the commandos from securing Zalakhan, with all its crucial insurgent paraphernalia, had failed dismally. The Taliban had, simply, been caught completely off guard.

As Canadian Leopard tanks ploughed a new way home clear of IED threats for L Company, and with K Company in their helicopters arriving back in Kandahar, 42's tired, dusty, thirsty, hungry and unshaven marines could reflect on an excellent start to their deployment.

With little respite from their exertions, the Commando reconfigured and rebombed in Kandahar and days later, on 14 October, flew north by C130 Hercules to Tarin Kowt, the capital of Orūzgān Province. A forty-six-vehicle logistic convoy led by the CRF had left Kandahar on the night of 15 October and by the evening of 16 October, all – men and logistics – were bedded down in what looked like a gypsy encampment on the edge of a dusty airfield outside the Dutch base, Holland. With the prospect of a helicopter assault into the green zone surrounding Mirabad, to the north of Orūzgān, and the prospect of some real fighting, the moral of Wilson's and Cantrill's marines was justifiably high.

The notorious Mirabad Valley, 120 kilometres north of Kandahar, was slightly narrower than that of Helmand, and had its own green zone of mostly wheat and maze, and some poppy fields. With miles of vines also planted across the green zone, the valley had been described as 'Tuscany without the wine', as the vines were for raisin production only. It was utterly beautiful, with a narrow river and a predominance of orchards filled with orange and lemon trees. But with complex terrain, narrow alleyways between plantations and log crossings no good for vehicles, it was also very difficult to manoeuvre in.

As Taliban activity in the valley had long been considered a threat to Tarin Kowt, it was an area of particular interest for more reasons than just the raisin crop. The insurgents were bringing ingredients for their bombs, plus ammunition and weapons, through the valleys then, having manufactured their IEDs, were taking them – to the great concern of the Dutch – into the capital.

Although the marines had already operated in an air assault role in the Panjvāī region, it had been with the Canadians, and so this was to be the first operation they carried out that was fully independent. Under Dutch command, there were also French infantry in the area, with whom a certain synchronisation was needed in order for each to own their 'personal' battle space. As a result the CO – Charlie Stickland – had to work out carefully his concept of operations and how to resupply across the differing terrain of open desert, the green zone and the hills rising out of it, and then desert again. He would also need to establish a number of over-watch positions covering the divide between the *dasht* and the green zone.

Operation Bor Baraki ('great thunder') was planned to find insurgent weapons, ordnance and equipment in the Mirabad Valley. It was an area that had never known conventional coalition forces and was thus considered another safe haven by the enemy, that is, until the British commandos announced that it was a job they could take on successfully.

Prior to *Bor Baraki*, the Australian forces had focused on the area, intent on seizing key insurgent leaders. As a result, the majority of fighting-age males had moved out, leaving a strange remaining demographic for L and K Companies to face before they could degrade Taliban stocks, weapons and explosives while disrupting Taliban command and control of the area. There were, however, enough AK-47-toting men around to satisfy the positive identification parameters, while most non-combatants, sensing that something was afoot, had also left.

The odds were very much in 42 Commando's favour, and the Commando Reconnaissance Force began dominating the valley from both the north and the south, before a rapid investment by L and K Companies in a helicopter assault into its very heart. Everyone, including the Dutch, had expected the marines to clear steadily from the west but that was not the commando way – the indirect approach was key – and the helicopter was instrumental to that victory.

The successful operation was initially kicked off when three fighting-age males were detained following a textbook surveillance by Recce Troop, who had been observing suspicious activity. As the Chinooks landed to begin the assault, these detainees were to provide instant and significant intelligence before they were moved up the detention chain.

Spreading out from their company and troop landing sites, the commandos then systemically and methodically began to secure, clear, search and exploit each area as they took it.

Fighting, and even simply just manoeuvring on foot when not in contact, in the dense undergrowth of the supremely fertile valley, was not easy, and required calm nerves and quiet minds. It was about as testing as it could be for junior leaders as they anxiously peered round and through thick hedges and tall crops before – using silent hand gestures – they moved their men onwards. It was a defenders' dream.

In the end, though, not a shot was fired. Anyone who gave 'probable cause' was detained. Huge quantities of explosives and weapons were found, while a good amount of time was spent talking to and understanding the local population.

To avoid setting a predictable pattern, the marines then conducted a good, old-fashioned yomp through the night, to reappear suddenly, quietly and in force further down the green zone. This led to more finds as L Company searched various objectives: IED equipment and cached IEDs, brand-new ordnance, weapons and ammunition were all unearthed by an increasingly inquisitive Company group. Full of confidence that night, the men of 42 Commando bounced in and out of the green zone and among the clusters of compounds and villages, each time finding more and talking more; the clear message to the local nationals being that security and peace of mind were being delivered.

Slowly, hesitantly, over the next few months, the local inhabitants began to grow accustomed to this large number of heavily armed and enthusiastic Royal Marines, who conducted themselves with restraint and cultural awareness. Lieutenant Colonel Stickland's Smiley Boys had it drummed into them that their message was that they came in peace, and with the concurrence of the Afghan Government. They were to sit and listen, take note of possible intelligence and soothe fears, while the medical and logistic teams working in two and threes slowly, patiently, diligently and politely, gained the locals' trust. With that trust came snippets of intelligence which, the marines would assure the Afghans, would be used only to their advantage and for continued security.

L Company's first significant find in *Operation Bor Baraki* had, at first, appeared to be only a mere handful of RPG warheads and 7.62mm ammunition dug into a compound floor, with further metallic objects unconvincingly buried beneath hay in an adjoining room. Fearing booby-traps, the counter-IED specialists had been called forward as the marines prepared themselves for the assessment that they had wasted their time with such a small find. However, and much to their delight, as the fourteenth 107mm rocket was unearthed from the floor from beneath the hay pile, they knew they had made the right call.

The clearance of other compounds within their immediate objective area also led to further weapons, ammunition and intelligence finds. L Company was quickly learning how to assess where and how key items might be hidden, while all the time making the best use of the Vallon metal detector.

The counter-IED team were kept busy bounding between such finds, and the Military Police ran out of evidence bags as well as the capacity to carry all the evidence being unearthed. A wheelbarrow was requisitioned by Major Cantrill but this quickly proved to be unworkable (much to its owner's relief, who was not happy seeing it used to cart loads of potentially unstable munitions across rough ground). Working in concert with K Company, the overall mission to disrupt was achieved, while the depletion of insurgent supplies was significant. One controlled explosion, which included 650 kilograms of homemade explosives, was so large even the counter-IED crew were surprised at the bang it made.

The men had done brilliantly, whether it was the men of Recce Troop and the Command Company watching patiently over their colleagues from the high ground; the lads in the companies themselves as they cleared rapidly from compound to compound and from village to village; the logisticians diving for cover as their parachute-delivered resupply floated to the ground; or the Viking drivers punching those supplies into the green zone. These last two were particularly crucial: successful operations don't just happen, and the resupply in the field of essential supplies could be conducted only by aviation (often as under-slung loads) to a desert laager. Air-dropped supplies, which included huge pallets of water, gave the Commando the agility and mobility it needed to over-match the Taliban.

The haul on *Bor Baraki* was so large that Stickland's Smiley Boys swiftly began to make a name for themselves across the international coalition as an eager bunch who could turn up a remarkable amount of information and hard evidence whenever they were out and about. Of almost more importance, also, they had gained a deep understanding of this previously uncharted territory, and their advice to their Dutch and Afghan security force partners on how to occupy and tackle the problems of the valley was to prove invaluable. Colonel Stickland wrote a mini-campaign plan for the Dutch commander, advising that what he needed to do was put the Dutch into the valley, sack the Chief of Police for corruption and incompetence, put the ANA four kilometres further into the valley and build them a base, and engage the Ghilzai (the area's local tribe) in *shura*. With that, he reckoned they would change the dynamic in the valley permanently.

Two and a half months after the operation, Stickland met the Dutch commander in Kandahar, who told him that they had done all that. The Chief of Police had gone, and in the last provincial *shura* there had been seventy Ghilzai elders. Even more satisfyingly, the valley had not seen an incident since.

After 12 days in the field, and following a short walk out of the green zone into the breathtaking scenery of the *dasht*, L Company returned to the Commando concentration area with a feeling of achievement of a job well done. In a sea-change to usual marine thinking, and grateful for those helicopters that they did have, the feeling was that the Commando would prefer to carry out more assaults by helicopter, despite 'Shanks's pony' being the Royal Marines' usual option (and sole guaranteed) form of travel.

Such was the success of their work, the men of 42 Commando were in no doubt that subsequent manoeuvres would be different from that time on, and that by heeding the lessons learned about clearance and search, the marines of Battle Group South could now face future operations across all the corners of southern Afghanistan with well-earned confidence.

Stickland's aim had been to act in true Commando style, using manoeuvrability, agility and superior strength to unhinge the insurgents and at each stage bite off and own sections of the valley. The Commando group had been prepared for all eventualities, from fighting to spending time and patience engaging the local population. On the day, the insurgents had not wanted to play, and this had afforded the marines the freedom to find the Taliban's arms and really understand the dynamics of the valley. Crucially, they had proved that conventional troops could operate in that key area, and had helped pave the way for future Dutch activity.

In an apparent endorsement of Stickland's views, Asadullah Hamdam, the Governor of Orūzgān, enhanced the influence of the Afghan authorities in the Mirabad Valley by hosting several *shura*s, while taking the opportunity to meet as many representatives from the valley's *kalay*s as possible. During these, he stressed the point being made all over Regional Command (South) that the eviction of the insurgents and the lessening of their influence offered a good starting point to build a more secure future. He told the local people that the stability and security in the region was not only the responsibility of the authorities and the Afghan national security forces, but also the duty of the population of the area themselves, adding that reporting insurgent activities was a good example of cooperation between the local population and the Afghan authorities.

As Regional Command (South)'s 'hired guns', 42 Commando (less J Company, which was employed at Lashkar Gah and Nad-e Ali) was required back to the Panjväï region in January 2009 to conduct a serious of operations in the *Shahi Tandar* ('royal storm') series of operations, north-west of Tarin Kowt.

An Anglo-Canadian blitz of Taliban compounds in the districts of Khakrez and Shah Wali Kot, which lie in the fertile green zone of the Tiri River, Shahi Tandar was another test of the commandos' ability to strike fast, unexpectedly and surely.

Between 7 and 9 January, Stickland's marines fought alongside Afghan national security forces; troops from the 3rd Battalion the Royal Canadian Regiment; and the 2-2 Infantry US Army. This phase lasted for three days, during which the force deliberately targeted the flow of weapons, explosives and IEDs used by the enemy.

Well before dawn on 7 January, the marines of K Company (L Company would have its chance a few days later) were up, breakfasted and checking last-minute details during their battle preparation routines.

Stickland's orders were for the assault to kick off with a dawn helicopter raid by K Company directly on to the first targeted compound then, supported by others from 42 Commando, the Canadians and the Americans, they would fight through the area seeking out Taliban fighters and their munitions. He had also decided that as the flight from Kandahar airfield would be only forty-five minutes, there was no need to stage through Camp Holland, to the south-west of Tarin Kowt.

As the mountains on the far horizon began to sharpen in silhouette against the dawn sky, the heavily laden marines filed across the tarmac and up the ramps into the waiting Chinooks. There was little space to settle in the canvas seats lining the darkened interior nor, for those standing, to ease the weight from their knees. Mild, good-natured cursing rippled through the fuselage:

'Mind your fucking pack, Smudge.'

'Sorry, Pony, didn't see you there.'

'Blind bugger. You'll be my first target, even before I get Terry in my sights.'

'Knowing how straight you shoot, I'll be bloody fine.'

'OoOOoo!'

With a small lurch the heavy-lift aircraft rose momentarily before gaining forward momentum. The noise increased, and further ribaldry was pointless. A cool wind swirled round the blacked-up faces of the marines and some shivered. The rising sun would soon sort that out.

The flight time meant that there was forty-five minutes for each marine to go through his mental checks: *Have I got everything? Where did the stripey say the emergency landing site was? Forget yesterday's passwords – remember today's. Why do they change them? Hope Jonno's leg's OK – he took a knock on the last op. Did I remember Susie's photo? Never been in action without it. Shit! Where the fuck have I left it …? Ah, thank fuck, got it. Whoah, look lively, the loadmaster's flapping like a loon, must be stand by to land. Hope the crab pilot faces this bloody machine in the direction the Boss asked for* … And so on.

The aircraft flared out to land. Sour-tasting dust hit the mouths and nostrils of the cramped men due to a sudden flush of hot air from the turbines. Standing marines clinging to sitting marines: *Now's not the time to fall over. Bloody difficult getting up again in this cramped space and with this load on … We're down.*

'Move! Move! Move!'

All right. All right. No need to shout. Can't wait to get off this effing machine.

'Thanks for the ride, mate.'

Even the loadmaster had to smile at that. 'Good luck, Royal.' He smiled again and meant it.

On target. On time. Good show. Well done, the RAF. That was hoofing; now where's my bloody section off to?

Despite training and rehearsal the men ran off the ramp. Never run, and never jump – the potential for injury, heavily laden and concentrating on other things – was too high. *Sod that, we can see the enemy compound. I'm not bloody walking into it.*

The marines ran straight towards the nearest cover, in the direction of the first compound. *No time to take stock. Exactly like the rehearsal of concept. Well done, Boss.*

The devastating speed of the initial assault out of the near-dark caught the Taliban completely by surprise. A few desultory rounds came down as the troops fanned out into their assault posture and advanced steadily towards their objectives, but otherwise resistance through the fifty or so compounds was limited, and eventually died away altogether, allowing the combined force to concentrate on the search phase of the operation.

This single phase of the *Shahi Tandar* series was as successful as the previous ones, and would be conducted again against other unsuspecting Taliban arsenals. Cleverly hidden in one compound, the marines discovered an impressive haul which included over 50 pressure plate IEDs, numerous IED components, over 100 detonators and 'det' cord for IEDs; plus six anti-personnel mines.

The operation had been a rapid, night-time helicopter assault directly on to the objective in the Zhari-Panjvāï district. It had been so swift and so unexpected that the enemy could offer only a very limited resistance.

Returning to Kandahar the way they came, and accompanied by eight suspected insurgents, the marines reflected on the operation's successes. Throughout, they had displayed a classic example of commando aggressiveness and boldness although, tragically, one Canadian soldier had been killed by an IED. A big price to pay.

The series of *Aabi Toorah* ('blue sword') operations that covered Nad-e Ali and the Fishhook reaches of the Helmand River had been set in motion by 42's *Operation Array* back in October. The aim of the series, as it was on so many operations in Helmand, Kandahar and Orūzgān, was to disrupt the Taliban across as wide an arc as possible with the forces available. To help out, 42 Commando was to operate, first, deep in the south of Task Force Helmand's area of operation and then almost immediately in Marjah and Trek Nawa, the troubled zone to the west of Lashkar Gah.

For the Fishhook phase of the operations – *Aabi Toorah 2B* – 42 Commando was ordered to conduct independent operations in areas and at times of their CO's – Lieutenant Colonel Charlie Stickland – choosing, based on daily intelligence that they themselves would acquire. To achieve this, they planned to cover over 1,000 kilometres across southern Helmand in Jackals and Vikings, after which they would then redeploy north, almost without a break, to conduct a helicopter assault into Marjah for the third phase, *Aabi Toorah 2C*.

The operation was to be unique among all *Herrick* deployments as the area had never been visited before by conventional troops: British forces, notably from the Royal Marines (whilst those from the army remained in Iraq as that conflict petered out), and American forces had patrolled the many small settlements and villages inhabited by Baloch tribesmen and 'narco' warlords, but this was, as an official MOD report noted, the first time such a size-able reconnaissance force had been into the area, particularly an area notorious as an ungoverned space that allowed freedom of movement for insurgent fighters, equipment and narcotics. Coupled with this, the operation was to be conducted far from help; further, in fact, from main headquarters and logistic support than any other since the ISAF troops first arrived in Afghanistan.

The Fishhook stretch of the Helmand River was so called due to a distinctive bend it had acquired over time: it flowed in a south-westerly direction from Kajaki, through Sangin, Gereshk and Lashkar Gah, and then curved sharply to the west, south of Garmsir, and it was this bend that was known as the Fishhook.

The wide expanse of desert both north and south of the river – a classic example of the Great Afghan Fuck All – was frequently traversed by Taliban moving up to the central and upper areas of Helmand from Pakistan, via the pass at Bahram Chah. An agricul-tural area, it was self-sufficient: the green zone was lush and produc-tive, and the farmers grew both barley and poppy so that they could have something both to eat and to get cash for. It was also an area that had been systemically destroyed by the Russians, with the result that the Taliban had had the good idea of explaining to the wary locals that the British troops were in fact the loathed Russians, speaking English as a deception.

Lieutenant Colonel Stickland's intent was to carry out a number of operation cycles with his Commando group by pushing into an area of his choice for two or three days, before withdrawing between ten and twenty miles back into the *dasht* to prepare for the next push into a different set of villages or *kalay*s. There would be a great deal of yomping: 42 was tasked to disrupt the insurgent, understand the area in every way and to identify key tribal and insurgent dynamics in order to inform and subsequently guide future Regional Command (South) HQ operations, and this was best achieved on foot.

It also meant there was far more to *Aabi Toorah* than simply an interest in the enemy's composition and possible future intentions. Within the battle group were specialist, non-kinetic effects teams – a smart name for a bright officer and a sergeant who 'did people'. Experts in the Baloch and Pashtun cultures, they would gain a better understanding of the region and its dynamics to pass back to Regional Command (South).

It was well known, too, that these actions were to be the precursor to a full-scale insertion by the United States Marine Corps' 4th Marine Expeditionary Brigade with 14,000 troops and 5,000 vehicles, some time in the near future, so the pressure was on for the intelligence to be accurate.

On 20 February, the Commando Reconnaissance Force, over twenty Jackals, set out from Camp Bastion. They were heading across the *dasht* for a reach of the Helmand River beyond the Fishhook, about 210 kilometres from their departure point and seventy kilometres north of the border with Pakistan.

The marines' navigation proved faultless over the three days, and the CRF flew across the desert's wide gravel plains – at times managing speeds of up to sixty kilometres per hour – to an LUP, where they arrived with time enough to spare for a few bonus hours in their sleeping bags.

Once on the move again, they began to draw closer to the green zone and, as the number of steep-sided wadis and deep defiles increased, the pace slowed.

On the way down, the Commando Reconnaissance Force had passed Garmsir and a pre-positioned emergency light surgical team should it be needed: the Commando's area of operating was beyond the radius of the MERT-equipped helicopters but not, thankfully, 'fast' air – the Royal Navy's Harriers and the American F18s – nor, of course, the ever-present and always welcome B-1B bombers.

One factor built into the Commando Reconnaissance Force's choice of each night's LUP was Major Adam Crawford's fear of snakes. Care was taken each evening to choose a spot least likely to harbour, in particular, the deadly saw-scaled viper.

Although Captain Orlando Rogers, Manoeuvre Support Group Commander, did not share his fear, the threat of being bitten by this particular viper was one to be taken seriously: one bite from a saw-scaled viper – they usually bite many times in the same attack – can mean death within two hours, and everyone was aware that with the lack of medical air cover, there would be no quick access to the venom's antidote.

One evening, when they had reached a night-time laager position supposedly clear of snakes, Rogers wandered off with his shovel for an evening 'squat'. While concentrating on other things, he suddenly saw, in the beam of his head torch, a pair of snake's eyes shining directly at him from the darkness. With Rogers knowing full well that saw-scales strike in offence as well as in defence, to his horror the snake had begun to slither forwards, making its distinctive 'sizzling' warning noise as it came. With inches to go, he managed to flick sand at its snout and it suddenly diverted, flowing quickly off into the darkness. An ensuing snake hunt with pistols chased it off, but everyone – particularly Adam Crawford – slept a little less soundly that night.

The Commando's intelligence officer had been honest enough to admit that he had no idea what 42 would find at their destination, which lay to the north of Zhrande Kalay. The area had never been visited by ISAF troops, and it was also tellingly close – about sixty kilometres of so – to the Pakistan border. Nor had he been aware of the soft sand, described by many as quicksand, that was to keep the Commando Reconnaissance Force's vehicle mechanic fully employed.

During a quick stop in Zhrande Kalay, the troops attempted to make up for this unusual lack of intelligence. Friendly or unfriendly, one or two civilians were willing to offer snippets of advice, starting with the fact that anyone with a sat-phone or a weapon in the locality would be Taliban – advice that was well documented and proved to be true and well attested. Zhrande Kalay itself was quiet, with not a mobile telephone in sight. Keen, however, not to overstay their welcome needlessly, the CRF travelled back out into the *dasht* to prepare for an insertion into Khan Neshin, their destination, twenty-five kilometres to the east along the river.

Khan Neshin was an ancient village that dated from Alexander the Great's era. Dominated by the ruins of a massive, square fort, its walls were over 170 metres in length. More recently, it had become home to a flourishing bazaar frequented by the Taliban.

The helicopter assault into the village had to be fitted together carefully if the enemy were to be caught unawares, and before they had time to take to the desert. It was also an area that needed investigating and understanding due to a long-planned, future US

Marine Corps surge into it, in order to break up what was certainly believed by the Taliban to be one of their safest strongholds, far as it was from ISAF activities. Being close to the Pakistan border made it an ideal staging post, command centre and logistics node, whose supply lines up from the south were used with impunity. It was high time that the Commandos carried out their mantra to 'find, disrupt and understand'.

For the past week or so, the two companies had been scouring a forty-mile stretch of the Helmand River's green zone, spending many hours in *shura*s, and drinking many cups of tea. So successful had these been, in fact, that when Stickland and his men left the area, it was with highly useful information for the coming action; intelligence that was also to assist the US Marine Corps when they became active in the area. But it was now time for more kinetic action.

On 27 February, twenty kilometres into the desert north-west of Khan Neshin, the marines of L and K Companies, plus sixty members of the ANA, were in their Viking-formed laager, ready for Chinooks that were due to arrive well before dawn. It was vital that the inhabitants of the proposed targets were kept in the dark, so the first contact was to be way to the west of the intended main target.

With the manoeuvre companies waiting out in the *dasht*, Adam Crawford's Commando Reconnaissance Force slipped as quietly as their Jackals' engines would allow through the dunes leading to the high ground, which overlooked the green zone, the ruined fort and, two kilometres further east, Khan Neshin itself. To help maintain as covert a start as possible to the operation, they knew there would be no signal to advance into an over-watch position and direct support; instead, the noise of the Chinooks bearing K Company as they approached, just before dawn, would be the command to advance. The CRF's timings had to coincide precisely with the arrival on the ground of L Company by foot and K Company by air.

'That's them coming in, sir!' A voice whispered in the dark as the first hint of the *whopah whopah* sound of twin rotors became distinct.

'Start up!'

With this, the Recce Troop careered off – no covert manoeuvrings needed now – racing forwards in a pre-dawn cavalry charge into their fire support positions that were on a ridge directly above the village, an over-watch that covered the three kilometres to the south-east. The remainder of the CRF moved a further one kilometre due south to line up along a similar ridge that overlooked the whole valley. Later, they would move into the valley to occupy the great fort.

Meanwhile, K Company was swooping towards the floor of the valley, their darkened aircraft – crews peering at the mottled-green landscape through night-vision goggles – curving to the west then swinging south to land in their trademark dust storms: on target.

With little chance of being engaged at that moment, the marines moved carefully off the stern ramp. As was usual, each man was carrying an average of forty-four kilograms, which included his Osprey body armour and water. Added to this, those with Javelin, and the troop sergeants who carried extra ammo, were carrying closer to fifty-five kilograms, and was why the marines were trained never to run off the back of a Chinook and certainly not to jump off it: a wound in combat was one thing, but a broken leg or twisted ankle was quite another matter, and one to be avoided at all costs.

The CRF had conducted preliminary moves even earlier in order to slip in to over-watch positions prior to first light. During this night infiltration, Lance Corporal Weaver had fallen hard into a deep irrigation ditch, badly dislocating his shoulder. As the whisper of his injury was passed over the command net the advance was briefly halted. A helicopter casevac so close to the objective would compromise the company's approach and warn the Taliban that 42 was moving against them in Khan Neshin. Aware of this, Weaver swallowed a mouthful of painkillers, redistributed some of his kit, bit down on his sweat rag and 'cracked on' in muted agony. He would not be lifted out until four kilometres and fourteen hours later after 42 Commando had defeated the core of the Taliban in the Khan Neshin DC.

Meanwhile K Company moved on from the landing site, finally reaching a position to the south of Khan Neshin before first light, just as L Company were successfully approaching their own line of departure prior to their sweep in to the village. That the Taliban were caught unaware of 42's approach was due to the stealth of every marine in the Commando. That the Taliban had not had time to arm and fully emplace their murderous IEDs was obvious later from the numerous pre-positioned bombs that led to the main bazaar along the approaches in to Khan Neshin.

Having approached 'so far' by Viking from their laager out in the *dasht*, the marines of L Company manoeuvred through the criss-crosses of irrigation canals as covertly as possible towards their start line, north of Khan Neshin. The plan was to infiltrate the village from the opposite direction that Kilo were taking by air in a classic pincer attack. Silently, tactically, L Company covered five or six kilometres but, in true Lima fashion, the marines had followed a tortuous route in darkness and across cultivated land, culminating in a water-obstacle crossing. A route no enemy would have expected than to have taken.

Until they reached it nobody was quite sure what form this final obstacle would take, other than that it was deep and wide. The leading troop brought up light aluminium ladders, used for scaling high compound walls, but they were not quite long enough. In the dark, a number of marines slipped into the water and thick mud, including the company's second in command, Captain Tom Dingwall, who had already earned the reputation for being fatally attracted to water. This latest episode, therefore, came as no surprise to the men, who found it difficult to maintain their covert presence while also suppressing their laughter.

Having negotiated this crossing the Fire Support Group moved off to the flank and on to the only patch of high ground actually in the valley, which overlooked Khan Neshin. Here they waited for four hours before Major Rich Cantrill passed an order to his three troop commanders. 'Saddle up, lads,' was whispered down the hastily prepared defence positions of low-built sangars and shallow shellscrapes his marines had scratched into the dry dust and gravel.

The rest of L Company's arrival, as planned, coincided with first light and the temporary occupation of a large derelict compound outside the village. Here 9 Troop, ever ferreting around, found a large stash of cannabis in a room that had been carefully lined in plastic. The professional nature of this store was one sign that the Taliban were probably *in situ* and in strength; as was hoped. Coupled with this was the more sinister indicator that the 'atmospherics' had changed: everyone had noticed the sudden absence, compared to earlier recces, of civilians. All the tracks and surrounding areas were empty, but following the recent engagements further down the valley this was, perhaps, not surprising. Now everyone – marine, corporal, SNCO and officer – knew that a battle – the longed-for scrap – was likely to happen.

At 0630, L Company began their final approach towards Khan Neshin, in order to secure a foothold from which more kinetic movement could take place. With 9 Troop leading, followed by 8 Troop, and with 7 Troop bringing up the rear with the Company HQ, all were suddenly halted in their tracks by a load explosion, followed by a black dust plume, which slowly drifted over the village. K Company were known to be in the area from which it had come.

'Bloody hell, that looks like an IED. Hope that's not one of our lads.'

'Don't stop, lads. Keep the momentum going.'

Later the Helmand Task Force log was to confirm what had happened:

OP AABI TOORAH – One suicide IED by a lone male in ditch acting suspiciously and staring nervously at friendly forces. Male then stood up and self detonated at Grid XYZ. Friendly call signs exploited area and now continuing on task but altering posture due to threat increase. No friendly forces casualties.

Once more the ground was littered with small, barely recognisable, body parts. Stepping over and around the obscene sight, the marines of K Company had continued their advance, the recent short silence quickly covered by ribald comments, black humour and not a few nervous laughs.

'Keep going lads. There's work to be done.'

'Very good, Sergeant Major! What's for scran this evening?'

L Company 'Main', covered by 7 Troop, now moved forwards to take up position within a group of compounds sandwiched between the river and fields. Here, the Company Sergeant Major, Ed Stout, began to take control. Sentries were placed, arcs of fire issued and a suitable helicopter landing site identified should a casevac aircraft be needed: the only choice was between the compounds, where it would be protected from enemy small-arms fire. However, the grid reference was merely noted and no further preparations made, as it was known from hard-won experience that if the enemy saw a large, open area being cleared it was for an obvious reason. The landing site would be secured when and if it was needed.

The company's main compound was split in two, with what looked like a long accommodation block with high walls dividing the sections. Being cautious, and always preferring the indirect route, entry was made though a mousehole that the Royal Engineers had blown, allowing positions to be taken up along one side of the compound.

Ordering a lance bombardier of the Royal Artillery on to the roof, along with a marine with a general purpose machine gun, Stout quickly got his other lads positioned as he wanted them.

The time was 0930. Happy with his dispositions, Stout moved indoors away from the increasingly hot sun. The normal routines for supporting a battle were up and running; the company second in command, a now-dry Captain Dingwall, was on radio watch; and the emergency helicopter landing site had been identified.

Stout's relief in the cool was to be short-lived, though, as the moment he stepped into the shade, the crack of a single shot made him step sharply back into cover. At first he thought it might have been a negligent discharge, but it was followed almost immediately by a far larger concentration of fire that started to hit the compound. Moving to the doorway, the CSM was greeted by a handful of rounds that struck the wall a few inches above his head.

Defying shouts to take cover, Stout began to walk calmly among the buildings to settle the men and give them encouragement where needed; including suggesting to a stubborn forward observation officer that he put his helmet back on. The man had to fold eventually under the CSM's incisive and perceptive ... shouting.

And then came the dreaded cry over the net: 'Man down!'

Stout lost no time: 'I want his Zap number. Quick as you can.'

'Zap number. Right. Wait out,' shouted back the voice.

It is a number that could be sent quickly over an insecure radio network, without giving away the casualty's real identity. Once he had the Zap code – the man down was Marine Wynn – and while waiting for the confirmation of location and a report on the extent of Wynn's injuries, the CSM prepared a clearing party to man the helicopter landing site. He also made a mental note of the time: the casualty had been shot at 0957.

Timings for the marine from now on were vital, the main objective being to keep within the 'golden' hour when the chances of survival were at their highest. Running through the back of Stout's mind was also the fact that the casualty had a brother who was a private in 2 Parachute Regiment. Stout's brother was a sergeant in the same battalion and he was therefore aware that Paratrooper Private Wynn had been casevac-ed from *Herrick 8* with an injury from a contact, so his thoughts were with the Wynn family, who were about to receive more bad news.

But Stout had other priorities at this moment. He assembled and took the landing site party to his chosen spot, pointing out the likely approach of the helicopter and potential hazards. Leaving Corporal Scott and Marine Cooper to protect the site, he went back to control Wynn's recovery to Company Main.

The battle was still in its early stages, but three rockets now suddenly screamed in and exploded: one landed on the empty side of the compound, one by the river and one close to the helicopter landing site. The obviously well-trained Taliban had the measure of what was up.

*

Stout now paused to take stock: the 'man down' was being dealt with; K Company was also in contact; and L Company was still clearing through their sectors of the village. All pretty normal business but, a few minutes later, yet another shout ran round the radio nets: 'Man down!'

This 'man down' location was closer to home: Ed Stout ran to the helicopter landing site to find Corporal Scott lying face down, his trousers round his ankles. Marine Cooper was kneeling over him, a Maglite torch in his teeth, staring intently at the man's buttocks. Before anyone could comment about this hardly being the right time or place for such behaviour, Cooper turned to the approaching first-aid team and in his distinctive Essex accent explained brightly, 'It's all right, Sergeant Major. I've found the hole.' Thankfully, it was only a shrapnel wound.

With the landing site now compromised and with the only identifiable alternatives also in range of small-arms fire, Plan B – extraction by Viking – was the preferred option for the two casualties. Returning to the main compound with Scott, Stout sent a nine-liner with his extraction plan. Scott's embarrassing but painful injury was described in the log:

M – shrapnel
I – buttock
S – bleeding
T – FFD (first field dressing) applied. Bleeding stopped

And Wynn's as:

M – GSW
I – Entry and exit wound to shoulder
S – Bleeding
T – FFD applied

Good news came with confirmation that Marine Wynn, although shot in the shoulder, was stable and in good spirits.

Major Cantrill, now making a new appreciation, pushed 8 Troop on to take out newly identified enemy positions. Things were hotting up: ordering 9 Troop to continue holding the ground they were occupying, the OC received a withering reply from a corporal: 'I can't move anyway, sir. I'm bloody pinned down as it is.' 8 Troop moved into an engagement that included an exchange of grenades at very close quarters, Apaches firing 30mm cannon at danger-close range, and a command course-standard troop attack followed by a section conducting an individual fire and movement assault. The position was finally taken by Corporal Nick Bonds – a physical training instructor.

The clearance of the objective – a small, fortified enemy compound – had been done by the book, and the arrival of the Vikings and the subsequent safe extraction of the two casualties left L Company in good spirits. Wishing to congratulate his boys, CSM Stout walked towards their newly captured position. Following the route of the assault, and moving past piles of empty cylinders and smoke-stained walls via grenade pins, kicked-down doors and enemy dead, it had been, he thought, a great day for the company.

One enemy fighter, attempting to escape L Company's ferocity, jumped into a canal only to slip downstream into Kilo's area, to the eternal surprise of a K Company sentry. The marine first saw a head and then – as their eyes met – a body armed with an AK47 appeared from the water. The two exchanged fire at a distance of five metres. The commando was quicker and the Taliban fighter fell back into the water to continue downstream in a bloody mess. An Apache then followed up with some 'rather close' 30mm rounds through the middle of the company's position to ensure the escapee was indeed dead.

*

The men of Recce Troop, still in over-watch on the ridge overlooking the *kalay*, had seen that K Company's first-light insertion had been met by 107mm rockets and, in support, had replied with machine gun fire of a variety of calibres, at the same time staking the claim of making the first casualties.

It was the catalyst the marines of K Company had been waiting for as they rolled in to clear areas of the village not occupied by L Company. The fighting had quickly intensified into bitter, close-quarter skirmishes as the commandos stormed enemy compounds; detonated mousehole charges; exchanged grenades with the enemy over compound walls; and launched dozens of individual assaults. The Taliban had replied to these with erratic attacks and hasty ambushes using automatic rifles and rocket-propelled grenades. Not once did they gain the upper hand.

'Hold tight, lads. Let Terry show himself.'

'There, Jim,' a marine pointed with his weapon. 'That fucking hole in the wall? Just seen a muzzle flash. That's one, all right.'

A double crack echoed off the walls of the narrow alley. Marines ducked instinctively but remained vigilant. Good training.

'Christ, that was close!'

'OK, I've pinged the bastard. Stand back.'

Marines crouched on one knee, their rifles on aim, pulled back tightly into shoulders. A burst of fire from the section's Minimi 5.56mm light machine gun echoed back as friendly fire ricocheted round the walls.

'Well done, mate. Keep going.'

'Where's the other section?'

'Going in now.'

'Bloody hell. They're moving fast.'

'Right. Move! Across the square. Take cover behind that fucking wall.'

'I'll cover you then it's your turn.'

Men fighting for their lives; men fighting for their oppos' lives. Not much thought of Queen and country. That comes later.

Right, it's me and my mate against Terry. Just one mistake and it's just my mate against Terry. Get it right first time. Remember Lympstone. Remember the training. Remember the 'thirty-miler' across Dartmoor. Remember the troop officers laughing out loud because they couldn't believe – any more than we could – that we had passed the Commando course! Remember the green beret presentation on the edge of the moor while our feet and backs were still knackered. Morale sky-high. Remember the next morning's hangover!

'RPG in-comer!'

'Bloody hell, that was too close.'

'Fuck me rigid, mate. Didn't see that one coming.' *Stop reminiscing – concentrate!*

'Come on, sunshine! Stop bloody dreaming. Move your effing arse or you'll be the subject of a nine-liner!'

While K Company moved through the area, another suicide bomber failed in his mission. Managing to kill only himself, his scorched, detached limbs ended up scattered over a thirty-metre radius. He had been sitting by the corner of a compound watching everything and everyone. He had looked suspicious but hadn't moved, and the marines had just kept well clear. Then, a sudden flash of light, an eardrum-blasting crack, a cloud of black and white dust and the horrific 'red mist' as eight pints of blood turned to spray. Everyone instinctively ducked although it was too late to avoid some unidentifiable bits of Taliban that came spinning across the gravel. As the smoke cleared and marines stood to shake their heads clean of dust, everyone scanned the ground around them – mercifully, there were no British casualties, no British body parts. Just Taliban – here, the enemy's whole left leg from thigh to toes; there, half a skull with one eye staring blankly along the sand; and, over there, testicles and a penis swinging obscenely off a bush.

'The heavenly virgins won't think much of that!'

Forget the patois of battle – the sights are far worse.

Systematically moving through the town, the fighting had intensified as the commandos closed on the main bazaar. Here, the insurgents attacked from urban firing points, including the numerous and much-dreaded murderholes, all the while exercising their usual disregard for the safety of civilians.

'Sir! Can you get that bloody Apache to earn his sodding keep?!'

'Call sign Ugly. In contact, Grid XYZ.'

A long, sustained sheet of 30mm cannon fire from the circling Apache overhead. The enemy's fire was instantly suppressed.

'Only joking. They're good lads, really.'

After checking there were no civilian casualties, the marines of K Company and their colleagues of the Afghan National Army could enter and clear yet more compounds: especially those closest to the bazaar where, almost by accident and in the middle of a dozen private fire fights, a section of marines stumbled upon a large quantity of raw opium, weapons, ammunition and RPGs. Chillingly, there were also two suicide vests primed and ready for use. They had already seen what those could do that day.

K Company's follow-up actions through alleyways and across small courtyards had been unremitting and fierce. Between them and L Company, they had inflicted a massive blow to the enemy, leaving six dead.

The Taliban was beaten and they knew it. From the moment they had heard the helicopters as the pilots flared out to land at the start of the dawn assault, they had known that they would lose.

Now they began to flee in disarray, the overmatch of the commandos simply too much for them.

With the insurgents fleeing from Khan Neshin, K Company, led by the ANA, began initiating reassurance and intelligence-gathering patrols, with the aim of reintroducing a legitimate security presence in the village.

This was always a difficult process, fraught with peril and the possibilities of misunderstandings: Stickland's men had to ask to search people's houses while trying to be balanced, and as sensitive as possible, while junior commanders had to negotiate a price per square metre with the elders before 'renting' a (usually occupied) compound as a base. Stickland always deployed his non-kinetic effects team at this point, and they helped smooth things considerably.

There is always a risk inherent with this sort of soldiering, but the marines found that living among the people – treating the people sensibly, rather than just walking into their houses – was to everyone's advantage. The Pashtun code demands hospitality, loyalty and blood-feud revenges, so it was necessary that the marines respected that code at every level. And by behaving 'normally' in an un-normal situation, there could often be the makings of a useful bond: with the marine ethos of choosing to carry kit over and above the need for a bivvie, the standard operating procedure was for the lads to go into a compound, agree compensation, establish force protection, establish communications and then beg a blanket. Stickland, however, always had a sneaking suspicion that his lads begged the blankets first …

Charlie Stickland had learned this method of operating from a lieutenant colonel in the Paras, who had perfected it and passed it on. Living in occupied compounds had many benefits, ranging from the protection perspective: the compound walls were very thick and the observation from the roofs was very good; to the fact that the set-up had the added bonus that the ISAF troops got to know the people. The marines drank *chai* with the owners of the compound where they were billeted, obtained a good feel for what was happening in the area, and became involved in the family's daily routine. It was influence through posture, and with that, an understanding of, and respect for, the *Pashtunwali* ('way of life') code, and the cultural sensitivities of the area. Before deploying to Helmand, the marines had been taught that the Afghans admire strength, and the ethos of an 'honourable warrior'. They had to be robust, firm and strong but culturally sensitive; never more so than in the Fishhook. Here were frightened people, led to believe by the Taliban that the British were solely there to remove their drugs (and therefore their means of earning cash), rape their women and generally behave as a marauding mass of looters.

Stickland joined his men for the first of many *shura*s held with local people, and further hearts and minds events were set up with the consent of the village elders: arrangements were made for the Commando dental team to establish a temporary clinic, thereby, it was hoped, proving to the population that ISAF troops were there to provide more than just force and security.

The locals began to engage with the marines and respond positively. When they offered honest feedback on their situation and a degree of hospitality it was, for a lot of the men, humbling. By acting decisively and fairly, it was felt that the ball had moved forward significantly in that area.

The final move of their long operation saw 42 Commando launch one more surprise move, to the western villages of Malakhan and Taghaz, just a few kilometres from the border of Nimruz Province. They travelled a lot of the way by either Jackal or Chinook but, approaching the target area it was, once again, old-fashioned yomping that produced the best results.

By now, the routine day-to-day operations were becoming familiar, and the marines and the ANA enjoyed 'day trips' to Divalak, a mere sixteen kilometres to the east, and to Pay Banadar, another three kilometres south-east across the river.

By the end of four weeks, very few areas in the Fishhook remained unvisited by the Commando, while the Manoeuvre Support Group even managed to alter the local agricultural system by driving through a fragile dyke and spectacularly bursting its banks. Despite this setback to the hearts and minds aspect of *Operation Aabi Toorah 2B*, the operation had been a success, finally coming to an end as the last of the ground troops rolled back to Bastion, gladly bidding goodbye to downtime boredom-induced pepper-sauce eating contests, days of breathing dust and Muscle Beach-style desert laagers.

42's proud commanding officer summed up *Operation Aabi Toorah 2B* as one that had been unique. Not only had it been in an as yet unvisited part of Afghanistan, but its helicopter assault directly against enemy positions had been tactics alien to the Taliban, who had been used to watching coalition forces approach from some away off, and often in daylight.

There had been an assumption that there would be a shortage of helicopters, but from 42 Commando's perspective that was not the case, although they were not always British aircraft they used (it was, also, equally true that 42's operations probably ensured that others went short). And it paid to remember that it was, ultimately, the men who won the battles, not the helicopters – on clearance or deliberate strike operations to gather intelligence, the marines had to be on foot, for maximum manoeuvrability. But the helicopters allowed flexibility in another way: by not working out of a fixed location, Stickland had the luxury of putting 500 marines into, for instance, a green zone objective in three waves within an hour to produce the required shock and dislocation.

Once the marines had conducted a helicopter assault, the Taliban would usually take about thirty-six hours to work out where the Commando was and what it was doing before trying to out-manoeuvre them with their Toyota Hilux wagons, motorbikes or on foot. Wise to this, the Commando would exploit an area for a maximum of forty-eight hours and then move out, either by helicopter or often by night, at which time the Commando Reconnaissance Force, in their Jackals, would shadow the main party from the flanks, sometimes from as far away as fifty kilometres.

Over the month that 42 was in the field, four operation cycles were conducted, each initiated by a Commando group aviation assault. They had yomped huge distances, searched every compound encountered, fought for a key town and drunk yet more tea in local *shura*s. They had been supported by their sandy echelon hidden in the desert fringes – the Commando Reconnaissance Force – and all the while re-supplied by well-coordinated aviation and C130 Hercules air drops.

Skinny, tired and bearded, the marines of 42 Commando returned to the delights – practical and emotional – of Camp Bastion, but within hours were propelling themselves yet again into rapid battle preparation for their final deployment – *Aabi Toorah 2C* – into an area just north of Marjah: effectively at the completely opposite end of the spectrum in terms of operational areas.

'Prepping' for the next operation before anyone stood down from the last was always a fundamental aspect of Colonel Stickland's standard operating procedures, with the quartermaster always having to plan two operations ahead of the one 42 Commando were conducting at the time.

It took about ninety hours' work to regenerate the Commando from coming in from an operation to going out on the next, and this time was no different. The moment the Commando arrived back at Bastion, the quartermaster met them in the vehicle park with all the specialists lined up and ready to go. Before anyone of any rank rested, all the weapons were to be mended, the radios recharged and the stores sorted out so that all the regeneration needed to be ready for the next operation was in place, and the Commando prepared. Throughout *Herrick 9*, Stickland held all his companies at two hours' notice to move. This was not standard, but it was vital.

An equally important part of the winding-down process after each operation was when Charlie Stickland would assemble his whole Commando to give a presentation that explained what they had done, and what they had achieved, in order that each company and sub-unit knew how they had all fitted in to the whole experience. It was always a much-appreciated finale to each operation.

In central Helmand, as opposed to places such as southern Kandahar, the commandos of 42 had found that the Taliban soldiers were more professional and better orchestrated than they had been during *Herrick 5*. Having fought them in the Fishhook, in central Helmand and in Panjväï, there was never any doubt that those in Helmand were at their most dangerous best.

CHAPTER SEVEN

MARJAH

Out of the dark came a question that was almost as adrenaline-inducing as had been the single rifle shot. A confused Taliban sentry stepped into the path of the CRF and demanded of the second in command whether or not he was a Soldier of Allah. The near-instant reply was an extraordinary Star Wars-style tracer display from the manoeuvre support group's wide array of light and heavy weapons.

For much of *Herrick 9*, as far as 42 Commando was concerned, if it wasn't Nad-e Ali that was on everyone's lips, then it was Marjah. In many respects the two went together. Now, with 42 Commando freshly returned from *Aabi Toorah 2B* deep in the south of Helmand, it was to be Marjah's turn for a night-time helicopter assault under the name of *Operation Aabi Toorah 2C*. The operation was set for the middle of March and with less than a month before flying home, the marines, with typical military humour, dubbed it *Operation Saturday Night Fever*, for it was now all about 'Stayin' Alive' …

This third and final phase of *Operation Aabi Toorah* had three main objectives: first, through rapid manoeuvre and build-up of combat power, it would surprise the enemy in their Marjah stronghold by disrupting their movement and planning; second, it would confuse the Taliban's situational awareness by hitting their fighting forces across a key, central location; and finally it was hoped the operation would allow the spread of legitimate Afghan governance, thus further enhancing the stability of the neighbouring districts of Nawa and Nad e-Ali – both of which had received the same treatment in the past few months from the rest of 3 Commando Brigade.

A secondary mission was also to create space in advance of the relief in place between 3 Commando Brigade and 19 Light Brigade, due to be completed with the transfer of authority on 10 April 2009. It was hoped that by sorting out the last of the bothersome district west of Lashkar Gah, the incoming brigade would be able to concentrate on Babaji. With too few men to hold captured ground the enemy would be back, but, it was hoped, to not quite the same troublesome strength.

Marjah had long been a region claimed by the insurgents as their heartland: a place they felt secure and where they could gather, equip and train their forces. It was where they stored weapons and explosives and it was where the links between the insurgents and the narcotics trade were at their strongest.

For most of 42 Commando's seven-month tour, the name of Marjah had cast a long shadow. Lurking in the wings as it always had done, when it was finally confirmed that Stickland's Smiley Boys were heading into this area with serious intent, it had brought a glimmer of relief, not from trepidation, but from waiting. For the men of the Commando Reconnaissance Force, who would be dropped into the back garden of the Taliban over two kilometres ahead of the main effort, for the Black Knights of Kilo who would initiate the break-in and then spearhead the advance, and for Lightning Lima who would fight through in support, it was what they had all been looking forward to doing.

Marjah, about ten kilometres south-west of Shin Kalay, is bounded by canals and an extremely confusing irrigation system. Stickland's plan for *Aabi Toorah 2C* – the subject of a precisely orchestrated rehearsal of concept at Camp Bastion – was to build on the successes of the full-scale helicopter assaults conducted so far. No one in the assault force would inserted by vehicles. Once the break-in had been achieved to this highly complex terrain, Viking Troop, Mortars and the Danish Mechanised Infantry Company would all move in to secure the 'Bridgehead'. The Commando's trusted Danish friends in their Leopard Tanks would also provide overwatch from the *dasht*.

The first insertion by CH47, USMC CH53, Sea Stallion and Royal Navy Sea King would be K Company, initiating the break-in by capturing the canal crossing to the north, which would allow for planned resupply, control of the area and the stopping of enemy reinforcements coming in from the *dasht*. K Company would have with them a lightweight infantry assault bridge to be used as a temporary foot crossing, which would be placed somewhere the Taliban would not have mined nor have covered by fire.

While K Company secured this crossing, the Commando Reconnaissance Force was to land in three Chinooks to form an outer cordon in the very centre of the Trek Nawa district, between Marjah and Nad-e Ali. To ensure that K Company's break-in was not compromised and to act as a simple diversion, they would over-fly the company's position to land a kilometre or so inside Trek Nawa itself. In depth, ahead of K Company and well within the urban areas, they were highly likely to be surrounded the moment they disembarked.

L Company, meanwhile, would land covertly in the desert to yomp, on orders – and with each man carrying upwards of 100 kilograms – the several kilometres to the canal crossing, from where they would control the roads, and prevent the enemy from fleeing with their arms and explosives. They would then support K Company in the steady clearing of Marjah.

There would follow a complex, combined arms battle, with a Danish battle group's Leopard 2 tanks, which had been used with great effect on *Operation Sond Chara* in December, moving from a laager way out in the desert to an over-watch position on high ground to the north and west. From here they would engage any dug-in enemy positions and opportunity targets. Vikings would also offer fire support and be available for emergency evacuation should the area be too hot for aircraft to land. Above, Apache and Cobra attack helicopters and a Bl-B bomber would keep watch and wait for instructions, while on the ground, 42 would be supported by GLMRs (guided rockets) and 105mm light guns from 29 Commando Regiment, Royal Artillery; 42 Commando's own mortars; plus a company from the Afghan National Army's 205 Hero Corps. The operation was expected to last five and a half days.

Although the odds were stacked in the ISAF's favour, the Taliban had two choices. Some felt that they would have a small go and then quickly retreat, leaving masses of hot intelligence, while others believed that the opposite would occur, in which case the worry would be compounded by the infantry component being made up of just three light companies with no vehicle back-up.

Typically, of course, the Taliban would choose the latter option. However, the outcome would prove extraordinary, with an estimated 100–120 insurgents killed in action and a similar number wounded. The ISAF was about to have an effect hitherto unknown in the Nad-e Ali, Marjah and Trek Nawa areas.

Operation *Aabi Toorah 2C* began on 19 March, with the pre-positioning in the *dasht* of the Danish Leopard 2s and the British Vikings, opposite the Trek Nawa district.

Major Adam Crawford wanted all 105 of his men to arrive on his chosen primary landing site at the same time and that meant using three aircraft in one wave. The evening before he had discussed with the pilots exactly where he had to be landed, the exact positioning of the aircraft to each other and the direction he wanted the stern ramps to be facing when his men disembarked. It was a good 'can do' conference, which allowed him to brief his lads comprehensively, even down to which way they should immediately turn on stepping off their Chinook's ramp. This was, in fact, crucial as Crawford was expecting to be in contact the moment they exited the helicopters. His lads knew that, too.

Shortly after K Company's helicopter package, supported by UK AH and USMC Cobras, had lifted off from Bastion at 0500 on 20 March, into the cool light of a Helmand dawn, the CRF took off in their three Chinooks.

It was an uneventful flight but, as they flew over the primary landing site that had been used by K Company's Sea Kings they were horrified – at least, those who could see out of the open stern ramp were – to see RPGs whizzing past the tail, apparently coming from their own designated landing site.

On the ground, K Company reported that they had observed muzzle flashes aimed at the Chinooks from just 200 metres from where they themselves had landed. Immediately after this news had been received, sustained anti-aircraft fire was aimed at the escorting Apaches, fired from the west.

Apache call signs Ugly 53 and Ugly 52 were swiftly on to the problem. They had good cause to be:

0545:   Anti-aircraft fire in the vicinity of attack helicopters. Observing; trying to positively identify further enemy firing points at Grid XYZ.

0552:   Positively identified heavy machine gun and enemy forces at Grid XYZ. HMG engaged with one Hellfire

> and 30mm. Several enemy forces killed in action. Secondary explosions. Wait for details of battle damage assessments. Engagements continuing.

Unbelievably, at the same time, an in-bound C-17 Globemaster was being engaged by a heavy-calibre weapon two miles south east of Camp Bastion. Although it was undoubtedly a coincidence, more reports were then received, this time telling of small-arms fire now coming from the CRF's intended landing site. The Taliban were intent on targeting the ISAF airpower.

With the Air Mission Commander – an Army Air Corps pilot – himself flying an Apache as escort, it was his decision, as the officer on the spot, to decide whether or not the landings should continue. Although the AAC officer had overall responsibility for the safety of the 'cabs', Crawford tried to impress upon him, over the radio, that he had to get them in there no matter what because if he didn't, the Commando plan would totally unravel before the operation had even started.

Luckily, Colonel Stickland was at that moment in Camp Bastion, standing alongside the AAC's commanding officer as they discussed the risks as the operation unfolded. He had only a few minutes before flying in to the desert landing site with L Company. As the CRF's primary landing site was obviously untenable, the decision to go for a secondary site was made instantly and the order passed to the pilots.

Crawford was not a little relieved. Their secondary landing site was about 600 metres further on, deeper in the green zone towards Marjah. They landed in the middle of a field, a site that had been discussed provisionally in planning, so everyone knew roughly where they were.

The Chinooks touched down almost together and in the same

relative position to each other as ordered. Needing no extra urging from the three loadmasters, the marines disembarked with as much speed as possible. It had been correctly assessed, thanks to the ubiquitous Information Exploitation Group's ISTAR operations, that the terrain's hundreds of ditches and cultivated ridges would be too constricting for resupply, so the marines had 'maxed out' on what they could carry – close to their own body weight – as well as filling dozens of sandbags with 7.62 ball and 5.56 linked ammunition. These were unceremonially dumped out of the Chinooks' stern ramps under the control of the troop sergeants. Once the aircraft had left and the dust had settled, everyone in the CRF HQ, including the OC, lugged these bags to a centralised dump. There were enough rounds in this cache to last all day, although some of the distribution would be later described as somewhat frantic.

As the brown-out dispersed into the dawn air, everyone took up in their pre-arranged all-round fire positions among the ditches, bushes, rubble and general agricultural detritus that was so familiar across Helmand's green zone.

Adam Crawford and his NCOs took stock. Less than two kilometres away to the north, K Company were noisily conducting a fire fight while Apache attack helicopters were engaging targets all over the place. Motorbikes and tractors full of fighting-age males were moving south, with ICOM chatter suggesting vast amounts of insurgent activity, including the knocking-through of murderholes.

With all this unlooked-for enemy activity, Crawford had to adjust his original plans. The CRF's main task was to act as a blocking screen to allow K Company to get on with clearing through their objectives without the enemy being able to bring in reinforcements. Now, in the half-light, he could make out the obvious routes that the enemy could take towards K Company's battle and despatched his sections accordingly.

Fanning outwards, the Recce Troop's leading section was immediately in contact but the ferocious exchange of small arms and grenades did not last long.

'Contact. Wait out!'

'Three of them! Watch to your left!'

'Christ, that was close! Cover me. I'm going to outflank the bastards.'

Pause. Small-arms fire cracked across the ditches, the AK-47 rounds making an eerie zipping sound as they scythed through the tall green grasses, then the unmistakable sound of a grenade exploding.

Another pause. Whose grenade? Ours or theirs?

'OK! I've got three dead.'

'Hoofing, mate!'

At the same time, within a hundred metres or so of their landing site, the CRF's Recce Team crept silently through a plantation towards a dirt track lined with ditches and hedges. Before they could crawl into their pre-planned ambush positions, however, a silver saloon car appeared, with three fairly senior-looking Taliban in it, speeding south. Even in the half-light it was obvious that they were armed and, if not up to much good at that precise moment, were about to be. If it wasn't engaged before it went past, it never would be.

No orders were needed. As the GPMG's rounds smashed through the silver bodywork, the car slewed to a sideways halt in a spray of gravel and sand. The three men leaped out clutching AK-47s and grenades, and dashed for the far-side ditch.

At a range of just ten metres, a close-range fire fight intensified until the Company Clerk, 'Essence' John O'Flagherty – who just happened to be with Recce that morning – decided that more offensive measures were the way forward. Creeping to within four metres of the enemy he initiated a grenade dual across the track.

'Grenade!' He shouted a warning to his mates to duck as he pulled the pin, stood for just long enough to judge the right direction and hurled his missile over-arm. Immediately he dropped to his knees, counting the seconds.

'In-comer!' Always a chilling shout. But, unwilling to take visual aim by popping their heads above their ditch's parapet, the Taliban's response exploded harmlessly in front of the marines.

'I'll get the bastards this time,' he muttered, then shouted, 'And another. Duck!'

The ensuing silence suggested that he had indeed 'got the bastards'.

Exhilarated at this encounter so soon after landing, the Recce Troop marines were in their element, especially as the Russian grenades had not been thrown very accurately, nor indeed very far. Dangerous? Yes, as the men knew, they only had to be unlucky once, but in this instance they knew that it was always going to be a one-sided engagement.

With the enemy dead it was time to finish off the contact. A second burst of Recce Troop machine gun fire into the stationary vehicle unexpectedly ignited explosives hidden in the saloon's boot, bringing the ambush to a close, with the unmourned loss of at least one senior Taliban commander.

An Apache flying overhead, unable to fire due to the risk of danger-close rounds, instead took a video of the incident; temporarily impotent, its very presence had helped to keep the Taliban's heads down.

0614:   Three enemy killed in action. Vehicle now cooking
        off with explosives.

But the team's work was far from done. A second dirt track ran close by, along which the occasional vehicle had also been driving. As they crept into a new position behind a small clump of bushes, they suddenly found themselves in another grenade-throwing competition with Taliban, just three metres away in a ditch, on the other side of the narrow road. After an exchange of GPMG and PKM fire from point-blank range, and after carrying out a textbook fire and manoeuvre to cover themselves as they moved, the team got even closer. Two grenades sent soaring over a wall into the insurgents' laps ended the scrap pretty definitively.

As Major Crawford and his mortar fire controller, Corporal Paul Tucker, orchestrated the clearance of compounds in their area, they moved towards the lead Recce Troop to better assess the need for support. Suddenly engaged by the enemy, the first either of them knew anything about it was when a burst of light machine gun fire tore up the gravel between them.

'Bugger this, sir,' a marine shouted from a ditch. 'Where're the bloody mortars?'

Corporal Tucker was on the case. Passing the coordinates to 42's mortar base-plate position, he ended his fire control orders with his stock-in-trade demand to, 'Rain the pain, mate. Out.'

The ICOM chatter also helped: 'The fools. They are targeting the wrong house. We're in the other one.'

Not for much longer.

Engaged by various weapon systems, including vehicle-borne SPG-9s that were fired from a variety of distances ranging from 150 to 500 metres, the CRF quickly realised the best way to deal with the threat was for the fire support team to call in everything from the GMLR – nicknamed the 'seventy kilometre sniper' – to the Apaches' Hellfire missiles. Corporal Tucker in particular was having a field day. At one stage his team, and others, were engaging with guns, mortars, GMLRS, 500- and 2,000-pound bombs from a B-1B Hellfire, and 30mm cannon from the Apaches, all in one go.

But it wasn't just the commandos' battle. By 0605, twenty-three separate enemy firing positions had been reported to Commando HQ and it was decided to offer some targets to the Apaches and Cobras. Others the marines kept to themselves, some openly declaring their enthusiasm for fighting a good enemy: 'Tell the bloody helos to back off. This is good sport!'

Morale was high as it was realised that the enemy action looked very much like an attempt to delay the ISAF, as they realised the Taliban were focusing on defending Marjah. To the north, K Company had been in almost continuous contact since landing while L Company, who by now had yomped in covertly from their desert landing site, were engaged south of the canal line. Each company gave and received mutual support, and all identified enemy positions surrounding 42 Commando were being systematically counter-attacked and cleared.

The three sub-units of the CRF and K and L Companies worked in concert, one always static in a position to cover the movement of the others. In this way, engaging the enemy both from the flanks and from within, the troops' movements were hidden from the enemy and, each time ISAF offensive action surprised them, worked by drawing them apart: the enemy's confused reactions enabled the ISAF and Afghan forces to pull them away from their command structures and defeat them in place.

A steady stream of wounded in action reports, however, were also filling the radio nets, all sent in accordance with standard NATO procedures:

0605:   Gunshot wound to left foot. Pulse 110. Resp 18. Little bleeding. First field dressing applied. Will be moved to the echelon via Viking.

0815:   Gunshot wound. Upper right leg. Entry wound bleeding. Pulse 72. Breathing OK. FFD and morphine given.

Once they had cleared each compound, ditch or defended position of insurgents, the marines secured the area and then took time to speak to and reassure the local people, who had had no time to make a run for it. Elderly and frightened men and women began peering warily around doors and over low walls, their faces blank with shock. Hesitantly, knowing the fighting for that particular moment was over, they began to gather in slowly increasing gaggles. In excited, emotional voices events were debated. Nervous glances were directed towards the combat-clad strangers but slowly, slowly, smiles replaced frowns. They were safe. The Taliban had gone. The Taliban, the men who tightly controlled the market forces in the area, had gone.

This pattern continued throughout that first, and longest, day. From the CRF's point of view, they did not need to move very far from the landing site as Crawford's men were doing exactly what the CO had asked of them: causing a diversion and preventing the Taliban from closing with the rest of the Commando. Meanwhile, K and L Companies were fighting their way with every weapon they possessed, including mortars and Javelin. Viking and Danish armour were used by the CO to give depth, fire support and plug holes in the counter-attack.

A situation report covering snippets of the first day emphasised the standard operating procedure of not firing at anything unless positively identified:

Attack helicopters were in support all day. Initially no enemy presence was observed but local nationals were seen leaving the area en masse from approx 0700.

The first real activity was observed south of friendly forces [these had been carrying out ISTAR duties] some five kilometres to the south-west of the break-in point, where a heavy machine gun and a bipod mount plus press-to-talk radios were seen being moved around. They mounted a compound roof to aim the weapon at the attack helicopters and were subsequently engaged with one Hellfire. Four enemy killed in action.

After midday, enemy activity increased considerably. All further engagements were in support of Lima Company who were receiving small-arms fire from up to twenty enemy positions. Ten 30 mm warning shots were fired into an open area [done in the hope of a reaction to which the marines could respond with more vigour].

At 1530 Lima Company still under accurate small-arms fire but no firing points positively identified. Eighty 30mm fired into clear area in the vicinity of one of the suspected firing points. Ten minutes later attack helicopters positively identified enemy positions in a compound engaging Lima Company. The compound was engaged with one Hellfire. At 1550 friendly forces came under small-arms fire again; 25 x 30mm fired into clear area as firing point could not be positively identified. One attack helicopter subsequently suffered a gun failure.

Enemy positively identified with weapons moving into a firing point compound at Grid XYZ. Engaged with one

Hellfire. Four enemy with light machines identified at Grid XYZ. Four enemy killed in action following 160 30mm and eight flechettes. At 1745 attack helicopters positively identified four enemy with light machine guns in the vicinity of Grid XYX. Engaged with 125 x 30mm. Four enemy killed in action.

From approx 1300 until 1830 attack helicopters were engaged with multiple effective RPG (airburst), anti-aircraft artillery airburst (AAA) and HMG. The AAA is assessed as 23mm, coming from the same area as yesterday – north of Marjah. Attack helicopters expended approx three Hellfire, eight flechettes and 400 30mm. Twelve enemy killed in action – estimate higher numbers in reality as the compound and tree lines are obscuring battle damage assessments.

Line and bearing of ICOM chatter indicates that enemy are in compounds at Grids XYZ and XYZ. They are shouting 'Allah Akbar' ['God is Great' in Arabic]. Other enemy firing positions are assessed to be Grids XYZ and XYZ. 105 artillery rounds on target but no effect.

Details are added to the starkly official reports recorded also at Bastion and Lashkar Gah, by individual marines' memory-snapshots of the battles they took part in.

One abiding recollection of some marines in the CRF was of 107mm rockets being fired horizontally at the wall of a compound within which men of the CRF were regrouping after a violent but successful skirmish. Even battle-experienced marines had never seen that before.

One marine was surprised by the anti-aircraft gun the Taliban used to engage the Apaches. It was surprisingly well hidden, and even with the helicopters coming back time and time again to get it, the pilots simply could not find it. The air above the battle was soon like a scene from the Second World War due to all the flak in the sky. A 'gift' from the Russians after they left the country in 1989, the Taliban regard their AAA – acquired on the black market – as their most prestigious weapon, along with their grenade launchers.

Another marine likened that first morning of the battle for Marjah to a Hollywood film. He had been in a ditch with the dirt kicking up immediately in front and all around him. The branches above his head were being mown off the trees and he had been amazed that no one was getting hit.

Adam Crawford's snapshot, finally, was perhaps more art-house cinema than Hollywood: forced to take cover in a noisome drainage ditch with fields of opium on either side, he had time to reflect – as he lay, three-quarters submerged – how pretty it was as the machine gun bullets zipped through the flowers above him, causing a shower of scarlet opium poppy petals to dust the filthy brown water around him. At that moment, he had realised, he would have willingly accepted a hefty dose of the stuff.

Towards evening, Stickland decided that the CRF needed to extract from their position overnight so that he could re-form the Commando. Crawford believed that he was fulfilling his command-ing officer's orders well: they had been very much surrounded all the time and under sustained attack throughout the day, which was what was needed to take the pressure off the other two companies. His inclination was, therefore, to stay where he was and get resupplied in his current position, but Stickland was adamant. Crawford's desire to stay put – and surrounded – was vetoed and he was ordered to move his men that night back into L and K Companies' fold. Fortune favours the brave and to enable Crawford's move the CO directed

L Company to push forward through the enemy line under cover of darkness to give the CRF less ground to cover and link up.

It proved to be a long night.

Some time after midnight and with Sergeant Foster, the navigation expert, in the lead – an unenviable position to be in but the price he so often paid due to his prowess as a navigator – the CRF began its move north. It was pitch-black, but Foster skilfully and silently skirted the more obvious-looking enemy compounds and ambush positions, all the while staring round intently through his night-vision goggles.

It was a common view that the Taliban, with their limited night-fighting abilities, did not like doing so in the dark; they preferred instead to hunker down for rest and food. But Marjah proved to be different. A whole Commando had infiltrated the very heart of their territory and by doing so had, almost purposefully, allowed itself to be engaged in a 360-degree battle – a tactic considered to be the best way to flush out the enemy and to tempt him into the open. The Taliban knew this and now proved that they could, when pushed, be as kinetic after sunset as they had been since that morning's sunrise. 'Terry' had to save not only his face, but his ground, too.

At 0415 on the morning of 21 March, two-thirds of the CRF was engaged from a compound that lay just 300 metres north-west of the landing site they had been fighting around: a compound that they had been watching throughout the day. Nobody was hurt by the initial crack of small-arms fire, but it sent adrenaline coursing through 105 suddenly rapidly beating hearts.

0415:   Contact. Grid XYZ. Returned fire with small arms. Friendly forces withdrew into cover. Am observing. Heat sources in vicinity of compound and tree line. Still receiving small arms fire. Intend engaging with Javelin.

With that nuisance dealt with, another new situation arose just as quickly: out of the dark came a question that was almost as adrenaline-inducing as had been the single rifle shot. A confused Taliban sentry stepped into the path of the CRF and demanded of the second in command whether or not he was a Soldier of Allah. The near-instant reply was an extraordinary *Star Wars*-style tracer display from the manoeuvre support group's wide array of light and heavy weapons.

As the CRF fired and manoeuvred out of the Taliban's killing zone, Marine Baines, who was operating the CRF's thermal-imaging equipment, was careful to stay with Sergeant Foster, who was still leading. Passing a chest-high mud wall, Baines suddenly grabbed Foster's arm. Through his night-vision goggles he had spotted three Taliban moving cautiously towards them just ten metres away, using an orchard for cover. The pair sank to their knees, followed quickly by the rest of the men behind them, each marine taking his cue from the one in front in a silent reverse form of a Mexican wave.

The Taliban, however, were just as alert and squeezed off the first shots. Within seconds, the marines out-gunned their adversaries with over-arching accuracy from small arms and grenades. Just as abruptly silence returned, while the dark hid from view the final spasms of three more dying insurgents.

If not a 'first', this contact by the enemy – in the dark and on the offensive – was an unexpected and unusual diversion from the Taliban's normal practice.

Adam Crawford was extremely worried about taking a casualty. At night and surrounded by enemy, it would have been extremely difficult to lift a casualty out, and certainly not within the 'golden' hour.

As far as they could tell, though, they had killed everyone in that recent ambush, yet the ICOM chatter hinted that there were others about. In the end, nothing further happened and with just 300 metres to go before linking up with L Company, the CRF was as home and dry as could be expected in that volatile situation.

Well before dawn, K Company moved south-west down the canal with support from the Leopard tanks on the higher ground at the edge of the desert. L Company was involved in this decisive phase – each company going firm while the other manoeuvred forwards in the classic infantry/tank cooperation mode.

After three hours of tense patrolling in the dark, the sun heralded not just another day but, inevitably, an almost immediate series of contacts. There was a lot of ordnance coming down, and lots of enemy about. L Company, who had a *Sun* reporter embedded with them, agreed not to mention it when the cameraman wet himself after a particularly intense barrage. They felt more strongly about the reporter's tendency to big himself up in his reports.

Having cleared through a patchwork of compounds near K Company, the Commando Reconnaissance Force was suddenly and unexpectedly engaged from four firing points, forcing the Recce Team to break contact and retreat by pepper-potting vigorously across open ground. It was a very close call that was brought to a swift end by an impressive display of ordnance from both mortars and 29 Commando Regiment's 105mm light guns out in the desert, which had been quickly called in by the gunners' forward observation officer. The CRF's fire support team also called in a B1-B bomber, which finally despatched the ambushers with a 2,000-pound bomb.

42 Commando's manoeuvre units continued their push south and west; never going firm, always on the move as the Taliban kept

coming at them: not in waves but by popping up in unexpected places. For those that experienced it, this period of *Aabi Toorah 2C* was regarded by most as the fiercest fighting they experienced throughout the operation.

Everything was working the way it should and the dynamics of the area were definitely beginning to change. By halfway through the afternoon of the third day, 22 March, the Brigade Commander took stock. Uppermost in his mind was that no one had been killed.

Late in the afternoon, while the Brigade staff deliberated, the Taliban finally broke contact for the last time. As the final shots echoed dully round the compounds, the mud walls absorbing the sound, the rifle companies knew that they had succeeded. They had finally cleared the area and created their own manoeuvre space. 42 Commando and its large array of supporting arms had produced the effect the Brigadier had demanded, and Charlie Stickland was also content that he had achieved his own aims. They had created the breathing space needed for the forthcoming transfer of authority, and the resulting calm in the area was one that would last for a couple of months. For such a short operation, it was an impressive result.

Calling a halt – 'Endex' – Stickland systematically withdrew his men, always keeping one foot on the ground, just in case: with the safety of a covering group, the sub-units yomped back to desert pick-up points from where Vikings, Sea Kings and Chinooks transported them to Camp Bastion for hot meals, cold showers and the inevitable soft drinks.

It was now the task of a Provincial Reconstruction Team to take over, and they set up an immediate *shura* held by District Governor Habibullah, at which the Nad e-Ali District Community Council and the Marjah elders discussed security and development. The aim, as always at these post-battle conferences, was to allow the people of Marjah to express their wishes for development through their representatives on the Community Council, and to share in the progress already enjoyed in Nad e-Ali itself since *Operation Sond Chara*.

Two weeks of post-operational administration now beckoned. All then that stood between Stickland's Smiley Boys and the rest of 3 Commando Brigade was a day's decompression and enforced banana boating in Cyprus, before they travelled home to see their families.

Lieutenant Colonel Al Litster, Chief of Operations for Task Force Helmand, summed up this phase as a very successful, deliberate joint operation that had demonstrated clearly to the enemy that the Task Force continued to operate where and when it chose. Marjah had previously been a safe haven for the enemy; ISAF had shattered that illusion, and more would follow. Friendly forces would continue to erode the capability and influence of the enemy, and enable the extension of legitimate governance throughout Helmand.

Operation Aabi Toorah 2C was an appropriate end to 42 Commando's tour, during which time, in the Regional Battle Group South, it had worked directly for Commander Regional Command (South) as an independent Commando group. Embedded in one of the regions' four Task Forces, the Commando had very much made its own luck, and exploited the opportunity to be truly expeditionary and manoeuvrist. For seven months, 42's core business had been deliberate helicopter-borne operations, whether to generate intelligence where there was none, or to influence or strike, to defeat and disrupt. These operations had generally been separated by a few days of frenetic rebalancing and battle procedures before they had tabbed off to another corner of southern Afghanistan. There was no doubt in Lieutenant Colonel

Charlie Stickland's mind that success had been delivered by the agility and robustness of his marines.

'Stayin' Alive' had been hummed incessantly throughout *Aabi Toorah 2C* and while some marines had suffered desperate, life-changing injuries, they had all indeed 'stayed alive'. Given the intensity of the fighting, this was quite an achievement, and one of which they were all justifiably proud.

EPILOGUE

'Our campaign in Afghanistan was not simply a fight against the Taliban. It was about supporting good Afghans to govern their own country so that we no longer have to fear a threat that emanates from it. Such a goal does not come without cost.'

EPILOGUE

BRIGADIER GORDON MESSENGER

COMMANDER 3RD COMMANDO BRIGADE

At the very beginning of his brigade's deployment on *Operation Herrick 9*, Gordon Messenger had laid down its mission:

> '*Task Force Helmand, in cooperation with the Provincial Reconstruction Team and Afghan national security forces, will conduct security and stabilisation operations in the focus areas and disrupt enemy forces' sanctuaries in order to deepen the Government of the Islamic Republic of Afghanistan's influence across Helmand.*'

At the transfer of responsibility to 19 Light Brigade on 10 April 2009, there was much to reflect upon, including the fact that any future brigade could continue to expect the same strategic surprises, tactical redirection, operational setbacks and unexpected strides forward that had typified *Herrick 9*.

It was also accepted that any progress made was not to be measured in terms of impact on the enemy but instead on the degree to which ISAF was able to reduce the political dynamics and geographical space in which that enemy thrived.

The core of ISAF work in Helmand is to provide the population with access to basic governance while building the people's confidence to accept that the presence of legitimate forces in sufficient strength will far outlast any insurgency. Where this was achieved during *Herrick 9*, an undoubted momentum in Afghan government was noticeable that, if replicated more widely, could one day provide the elusive tipping point that is sought. Conversely, where it was not possible to provide sufficient resources, antipathy and ambivalence towards the Government were prevalent.

During *Herrick 9*, highly successful initiatives such as the wheat seed distribution, voter registration and numerous, positive *shura*s allowed local Afghan governance to reach out to the population on a scale that was unheard-of a year before. As this is at the heart of the security effort, any future composition of the 'on duty' brigade and its tactics should, it is believed, be configured accordingly.

By the end of the tour it was certainly felt that a better understanding of the scale and nature of the challenge that confronts ISAF in central Helmand as a whole had been achieved. But there were serious caveats to any thoughts for continuing success as the Brigadier was to state: '*Following the necessary clearance operations, this will involve the additional allocation of two battle groups to hold Nawa, Babaji and Marjah. Until ISAF can generate these forces, our only option is to conduct a series of containment operations designed to orientate the enemy defensively, the effectiveness of which will inevitably dwindle over time. Elsewhere in Helmand, I believe we can afford to maintain economy of force operations in Musa Qaleh and Sangin, provided security and governance capacity is not seen to regress.*'

Behind this view was the knowledge that it had been necessary to form, at Brigade level, a new battle group (Centre South) and at Commando level – 42 Commando – a new manoeuvre company: the Commando Reconnaissance Force. Added to this was 45 Commando's experience of taking over the same commitments as its predecessors yet with, effectively, 300 fewer men, which was another case in point for the need of more resources.

There were other caveats too. The current practice of rotating whole Brigades every six months was considered an impediment to the campaign's progress. It takes time for each new unit to develop the necessary, deep knowledge of Helmand needed to operate effectively for the common good of the locals: just when the level of understanding has been achieved it is usually time to go home. Local Afghans are known to be frustrated by this constant turnover.

Another consideration is the pre-deployment training, as the Brigadier was also keen to stress: '*We need to train for counter-insurgency, not for war. Nothing I have seen or heard on this tour has involved anything other than well-disciplined, proportionate and legitimate use of firepower – the product of a successful and well-tailored training regime – but our mindset is still too kinetically focused for the real challenge we face, and we risk alienating the population as a result. Our training must include greater cultural awareness down to the lowest level, including basic language skills for all. More emphasis must be placed on effective and nuanced judgemental training at all levels, and on the intelligent, de-escalatory use of force.*'

Nevertheless, all felt that the campaign in Helmand was heading in the right direction (although always vulnerable to reversal). One widely acknowledged highlight of the tour was the increased integration of civilian and military organisations, with the expectation that the PRTs will, in due course, assume an even greater international influence while retaining their status as the focal point for all provincial engagement.

Another success, thanks to the dedication and patience of 1 Rifles, was the great stride forward made by the Afghan National Army's 3rd Brigade of the 205th Corps, although there remained a fear that such progress might begin to lessen, particularly with the projected shortfall of Afghan security forces in the south.

Such progress, though, comes at a significant cost in human lives and, tragically, at a rate that everyone must expect to continue.

The more formal words of 45 Commando's post-operational report by Lieutenant Colonel Jim Morris complete the story and are a tribute to all that 3 Commando Brigade achieved across Helmand Province. The same could well have been written by the commanding officers of all the other units that made up *Operation Herrick 9*:

> '*Our deployment has seen genuine, tangible and sustainable progress in the delivery of security and local governance to the Afghan population within the area of operations. Indeed, there are now eleven schools that have opened since our arrival in Helmand; families are moving back into southern Sangin; the bazaar is thriving; and we have secured an area in the centre of town that is about to receive a new Afghan health clinic, school and Government offices. We have put in place the groundwork for representatives from a wider range of Afghan ministries to join the District Governor in Sangin, thereby extending the influence of Afghanistan's government further into the area. We also enabled the conduct of voter registration in preparation for elections later in the year and the number of local Afghans who came to register exceeded even the most optimistic predictions.*

'This demonstrable success has come at a cost, and as we return we remember and pay great tribute to those who have made the ultimate sacrifice and many more who have been wounded in battle during the course of the deployment. Enduring and sustainable progress was certainly made but it was gritty, slow and dangerous work. The determination to succeed and the courage, intelligence and cheerfulness displayed by the young men and women was remarkable.

'As we leave theatre there are clear signs of tangible and enduring progress across the area of operations and it is hoped that we have developed the campaign further, building on the work of our predecessors by establishing a degree of much-needed campaign continuity and creating the firm foundations upon which our successors can now build.'

Nor should the equally pressed marines, sailors, soldiers and air personnel of the other units and corps be forgotten.

The ubiquitous Information Exploitation Group had been involved in every action, large and small, through the constant and intelligent use of its wide and often secret range of ISTAR assets. Without the IX Group there would have been few, if any, successes.

It was the patient and professional soldiers of 1 Rifles whose task of mentoring and training the ANA who had probably the most important long-term success of all, while the Danish Battle Groups' Leopard tanks, so often out of sight but always on call, reversed many a sticky situation.

Similarly, the noble men (and women) of the Royal Artillery and Royal Engineers: the gunners and sappers deserve every possible form of praise, as do the drivers, mechanics and logisticians of the Commando Logistic Regiment, without whom nobody would eat, drink or fire a weapon.

The attachment of C Company the Princess of Wales's Royal Regiment with its Reconnaissance Platoon and mortars was invaluable. So, too, were 2 Battalion the Royal Gurkha Rifles, and the exemplary technicians of the 3rd Battalion the Royal Electrical and Mechanical Engineers. And, of course, in the increasingly cyber-aware theatres of modern warfare, the Royal Corps of Signals must be fully acknowledged and appreciated.

The medical teams of all services, including the Royal Marines Band Service, should always be singled out for supreme admiration. There are so many lives saved, and so many wounded combatants sent home in a far better state than their initial injuries might have indicated.

The base staff – especially the cooks and postmen (and women) – at Camp Bastion must never be forgotten.

Pilots and aircrew, so often accused of being divorced from the heat, cold, wet, fear, horrific sights, smells and ear-splitting sounds of battle in their Harriers, Sea Kings, Lynx, Chinooks, Apaches, Cobras, Sea Stallions, Black Hawks, B-1B bombers and the rest saved the day, and lives, as routine.

Shabash to them all.

OPERATION HERRICK NINE
HONOURS AND AWARDS

(Including those who, although not in Helmand Province, nevertheless worked in support of the Brigade throughout *Herrick Nine*)

Bar to Distinguished Service Order (DSO)
Brigadier Gordon Kenneth MESSENGER DSO OBE, Royal Marines

Distinguished Service Order (DSO)
Lieutenant Colonel James Andrew John MORRIS, Royal Marines
Lieutenant Colonel Joseph Desmond CAVANAGH, The Rifles

Officer of the Order of the British Empire (OBE)
Lieutenant Colonel Alan LITSTER MBE, Royal Marines
Lieutenant Colonel Charles Richard STICKLAND, Royal Marines
Wing Commander James Peter GRINDLAY, Royal Air Force
Brigadier Mark Julian HALLAS ADC, Late Intelligence Corps
Lieutenant Colonel Alan Stewart RICHMOND, 1st The Queen's Dragoon Guards

Member of the Order of the British Empire (MBE)
Warrant Officer Class 2 Kevin John CHEESEMAN, Royal Marines
Major Adam Timothy Stephen CRAWFORD, Royal Marines
Surgeon Lieutenant Henry DOWLEN, Royal Navy
Major Tristan HARRIS, Royal Marines
Major Ross Walker PRESTON, Royal Marines
Captain Brian Headrige CALDER, Royal Logistic Corps
Warrant Officer Class 2 Simon James William HALL, Royal
    Logistic Corps
Major Michael George TAYLOR, Royal Logistic Corps
Major Nathan Charles WEBBER, Royal Regiment of Artillery
Major Colin Neil WITHERS, The Parachute Regiment Territorial
    Army

Royal Red Cross
Lieutenant Commander Alison Jayne HOFMAN ARRC, Queen
    Alexandra's Royal Navy Nursing Service

Conspicuous Gallantry Cross (CGC)
Marine James MALONE, Royal Marines
Marine Steven NETHERY, Royal Marines

Military Cross (MC)
Marine Samuel ALEXANDER, Royal Marines
Corporal John BALLANCE, Royal Marines
Corporal Richard BATEMAN, Royal Marines
Major Richard John CANTRILL, Royal Marines
Sergeant Noel Gerard CONNOLLY, Royal Marines
Sergeant Andrew LEAVER, Royal Marines
Medical Assistant Class 1 Kate Louise NESBITT, Royal Navy

Lieutenant James Philip ADAMSON, The Royal Regiment of Scotland

Captain Xavier Luke GRIFFIN, The Parachute Regiment Territorial Army

Gunner Grant Michael GUY, Royal Regiment of Artillery

Lance Corporal Sean Michael KEENAN, The Rifles

Warrant Officer Class 2 Benjamin Llewellyn KELLY, The Princess of Wales's Royal Regiment

Acting Serjeant Mark John POWIS, The Rifles

Lance Corporal Gajendra RAI, The Royal Gurkha Rifles

Sergeant Torben Erik George SORENSEN, The Princess of Wales's Royal Regiment

Lance Corporal Colin James SPOONER, The Princess of Wales's Royal Regiment

Distinguished Flying Cross (DFC)

Lieutenant Commander Gavin Ian SIMMONITE, Royal Navy

Queen's Gallantry Medal (QGM)

Warrant Officer Class 2 Colin Robert George GRANT, Royal Logistic Corps

Acting Warrant Officer Class 2 John Gareth LESTER, Royal Logistic Corps

Staff Sergeant Anthony David SPAMER, Corps of Royal Engineers

Mention in Despatches (MiD)

Corporal Russell Howard COLES, Royal Marines

Lance Corporal Steven Daniel FYFE, Royal Marines

Leading Medical Assistant Richard HOGBEN, Royal Navy

Acting Corporal Adam MABROUK, Royal Marines

Corporal Samuel Joseph McCORMICK, Royal Marines

Sergeant James Ian MELHUISH, Royal Marines
Marine David George MIDDLEMAS, Royal Marines
Corporal Scott MUIR, Royal Marines
Marine Iain Andrew PENROSE, Royal Marines
Corporal Mathew SILCOCK, Royal Marines
Major Nigel John Powell SOMERVILLE MBE, Royal Marines
Acting Sergeant Jason Paul WALKER, Royal Marines
Major Andrew Patrick Leonard WATKINS, Royal Marines
Corporal Andrew WATT, Royal Marines
Warrant Officer Class 2 Adrian WEBB, Royal Marines
Corporal Thomas WEBSTER, Royal Marines
Corporal Edward James WINSLOW, Royal Marines
Marine Mark ADAMS, Royal Marines
Captain Rupert Timothy ANDERSON, The Royal Gurkha Rifles
Craftsman Martin BANKS, Corps of Royal Electrical and
    Mechanical Engineers
Captain Alex John BURGESS MC, The Princess of Wales's Royal
    Regiment
Captain James Alexander CARROLL, 1st The Queen's Dragoon
    Guards
Rifleman Darshan CHAMLING RAI, The Royal Gurkha Rifles
Major Ross James DAINES, The Royal Gurkha Rifles
Rifleman Manju GURUNG, The Royal Gurkha Rifles
Sapper Peter Joseph HEMBRY, Corps of Royal Engineers
Captain Daniel Paul HOLLOWAY, The Yorkshire Regiment
Lieutenant Edward Richard HUNTER, The Princess of Wales's
    Royal Regiment
Captain Pieter Alexander LAHORGUE, The Royal Logistic Corps
Rifleman Christopher Philip LOCKE, The Rifles
Captain James Edward Michael McCARTHY, The Rifles
Rifleman Stuart Winston NASH, The Rifles (Killed in Action)

Acting Serjeant Nicholas James NESBIT, The Rifles
Bombardier Paul James QUAINTANCE, Royal Regiment of
  Artillery
Corporal Basanta RAI, The Royal Gurkha Rifles
Sergeant Bikash RAI, The Royal Gurkha Rifles
Corporal Bishwahang RAI, The Royal Gurkha Rifles
Rifleman Gopal RAI, The Royal Gurkha Rifles
Lieutenant Drew Steven REED, The Princess of Wales's Royal
  Regiment
Captain Thomas William ROSE, The Royal Gurkha Rifles
Sergeant Brendan John VANNER, Corps of Royal Engineers
Second Lieutenant Miles Robert WATT, The Princess of Wales's
  Royal Regiment
Acting Captain Iwan Rhys WILLIAMS, The Rifles

Queen's Commendation for Bravery (QCB)
Gunner Sam JOHNSON, Royal Regiment of Artillery

Queen's Commendation for Valuable Service (QCVS)
Commander Gail Margaret AXON, Royal Navy
Sergeant Derrin Mark CANTERBURY, Royal Marines
Sergeant James HORROCKS, Royal Marines
Acting Warrant Officer Class 2 Garry MASON, Royal Marines
Colonel Andrew Thomas Westenberg MAYNARD, Royal Marines
Corporal Elvet Llewellyn WILLIAMS, Royal Marines
Major Julian Graham WILSON, Royal Marines
Major Thomas Howard BEWICK, The Rifles
Major Simon Peter BROWNING, The Royal Logistic Corps
Lieutenant Colonel Christopher Bernard DARBY, The Royal
  Gurkha Rifles
Captain James Andrew HADFIELD, The Rifles

Staff Sergeant John Sebastian REEVES, Corps of Royal Electrical and Mechanical Engineers

Captain Richard James SMITH, The Parachute Regiment Territorial Army

Corporal Mohan THAPA, The Queen's Gurkha Engineers

GLOSSARY

Of necessity this list is lengthy: fuller descriptions of many of these entries will be found in the text.

16 AAB	16 Air Assault Brigade
ADZ	Afghan Development Zone
Afghan bee	incoming small arms – because of the distinctive noise they make
AGS 17	Soviet 30 mm grenade launcher with a range of 1,700 metres
AH	attack helicopter
AI	Armoured Infantry (Estonian)
AK 47	Kalashnikov 7.62mm (short) assault rifle with a range of 800 metres
ANA	Afghan National Army
ANAP	Afghan National Auxiliary Police
ANCOP	Afghan National Civil Order Police
ANP	Afghan National Police
AO	Area of Operations
Apache	attack helicopter (AH-64) flown by the Army Air Corps
ASC	Armoured Support Company (Vikings). Royal Marines
ASNF	Afghan Special Narcotics Force
ATV(P)	Viking all-terrain vehicle (protected)

B-1B	USAF variable-wing, long-range 'Lancer' bomber capable of operating above 30,000 feet delivering a wide range of guided and unguided munitions
Banyan	picnic on a tropical beach – beneath a banyan tree
Barma	operation conducted on foot ahead of a patrol to detect IEDs
Barmine	rectangular anti-tank mine used in Helmand to blast an entry through an obstruction
BDA	Battle Damage Assessment
Bde	Brigade: commanded by a Brigadier
BG	Battle Group: Commando- or Battalion-sized all-arms group under a single command.
Blind	unexploded shell or grenade – a dud
Bootneck	slang for a Royal Marine
BPT	Brigade Patrol Troop, part of BRF
BRF	Brigade Reconnaissance Force
C2	command and control
.5 cal	.5 inch, belt-fed, Browning M2 heavy machine gun carried by WMIK and Chinook. Effective range 1,850 metres
Cdo	Royal Marines Commando of about 700 men (reduced to 555 for H5)
3 Cdo Bde	3 Commando Brigade, Royal Marines
CF	Coalition Forces
CGC	Conspicuous Gallantry Cross
CH47	Chinook helicopter with a 'normal' payload of forty fully equipped soldiers
CIMIC	Civil Military Co-operation
CNIK	counter-narcotics infantry *kandak*
CO	Commanding Officer of a commando/battalion-sized unit: lieutenant- colonel
COMBRITFOR	Commander British Forces

CONOPS	Concept of Operations
COP	Combat Outpost
Coy	Company: up to four troops including a Fire Support Troop. Four in a commando, each up to 150 men commanded by a major
CP	check point
CP	Command Post
CQMS	Company Quartermaster Sergeant
CRF	Commando Reconnaissance Force
CSG	Command Support Group Royal Marines, now renamed 30 Commando Information Exploitation Group Royal Marines
CSM	Company Sergeant Major
CTCRM	Commando Training Centre Royal Marines
CTR	close target recognition
CVR(T)	Combat vehicle reconnaissance (tracked)
Dasht	Pashtun for desert
DC	District Centre or Compound
DCC	District Co-ordination Centre
DfID	Department for International Development
Dicking	Expression from Northern Ireland: to be watched and reported by non-aligned – but usually hostile – civilians
DIT	Development and Influence Team
Dits	Stories – usually tall stories of derring-do
DSO	Distinguished Service Order
Dushka	Soviet 12.7mm equivalent of the .50 cal
ECM	electronic counter measures
EF	enemy forces
E-HLS	emergency helicopter landing site
ELINT	electronic intelligence
EOD	explosive ordnance disposal
E-WMIK	enhanced WMIK

F-15	USAF fighter, ground attack aircraft
F-18	USAF fighter, ground attack aircraft
FAC	Forward Air Controller
FARP	Forward Air Refuelling Point
FCO	Foreign and Commonwealth Office
FF	friendly forces
FFD	first field dressing
FLET	forward line of enemy troops
FLOT	forward line of own troops
FOB	forward operating base
FOO	Forward Observation Officer, Royal Artillery
FSCC	Fire Support Coordination Centre
FSG	Fire Support Group
FST	Fire Support Team
Galley	Royal Navy/Marine term for kitchen
GIRoA	Government of the Islamic Republic of Afghanistan
GMG	grenade machine gun firing 40mm grenades with a range up to two kilometres
GoA	Government of Afghanistan
GPMG	7.62mm, belt-fed general purpose machine gun. Range up to 1,800 metres
GPS	global positioning system
Green Eyes	unmanned aerial vehicle
H4	*Operation Herrick 4* (16 Air Assault Brigade)
H5	*Operation Herrick 5* (3 Commando Brigade, Royal Marines)
H6	*Operation Herrick 6* (12 Mechanised Brigade)
H7	*Operation Herrick 7* (52 Infantry Brigade)
H8	*Operation Herrick 8* (16 Air Assault Brigade)
H9	*Operation Herrick 9* (3 Commando Brigade, Royal Marines)

H10	*Operation Herrick 10* (19 Light Brigade)
H11	*Operation Herrick 11* (11 Light Brigade)
Harrier GR9	RAF and Royal Navy vertical take-off, ground attack aircraft
HE	high explosive
Heads	Royal Navy/Marine term for latrines
Hellfire	air-to-ground anti-armour, fire and forget missile system used by AH against fortifications and compounds. Range between 500 metres and eight kilometres
Hexamine	small, solid fuel tablets for heating water and food
HICOIN	High Intensity Counter Insurgency
Hilux	Toyota, hybrid, four-wheel-drive saloon with a 'pick-up', open back: often containing Taliban in the front and a mortar base-plate and other heavy weapons in the back
HLS	helicopter landing site
HMG	heavy machine gun
Honking	Royal Marines slang, antonym to hoofing
Hoofing	Royal Marines slang for 'excellent'
HRF	Helmand Reserve Force
HRM	Helmand Road Map
HSSG	Helmand Security Steering Group
HUMINT	Human Intelligence
HVT	high-value target
ICOM	hand-held VHF used by the Taliban
IED	improvised explosive device
ILAW	L21A1: 84mm, one-shot, unguided anti-armour weapon. Range 300 metres
IRT	Immediate Response Team
ISAF	International Security Assistance Force
ISF	Indigenous Security Forces

ISTAR	information, surveillance, target acquisition and reconnaissance
IX	information exploitation
Jackal	4 x 4 patrol vehicle capable of carrying combinations of .50 calibre (12.7mm) machine guns, Heckler & Koch 40mm grenade launchers and general purpose machine guns. Its hull is shaped to minimise the effects of a mine strike
JARIC	Joint Air Intelligence Reconnaissance Centre
Javelin	fire and forget, anti-tank missile system used in the 'bunker-busting'mode
JDAM	Joint Direct Attack Munition
JDCC	Joint District Coordination Centre
Jingly truck	Brightly coloured, heavily ornamented, Afghan civilian, heavy-lift lorries
JNCO	junior non-commissioned officer: lance corporal and corporal
JOC	Joint Operations Centre: providing the infrastructure for 42 Commando's Battle Group Headquarters
JPCC	Joint Provincial Coordination Centre
JTAC	Joint Terminal Attack Controller
Kandak	Afghan battalion
KIA	Killed in Action
.338 LRR	sniper rifle. Accurate range 1,100 metres
LD	Light Dragoons
LD	Line of Departure: start line for a deliberate attack
LMG	7.62mm magazine-fed light machine gun. Range 1,100 metres (Taliban)
LMG	Minimi, light machine gun (UK)
LUP	lying up position

MC	Military Cross
MCF	Military Construction Force
MERT	medical emergency response team
MFC	Mortar Fire Controller
MiD	Mention in Despatches
Minimi	5.56mm, belt-fed light machine gun. Range 1,000 metres (UK)
Mne	Marine
MOB	Main Operating Base
MoD	Ministry of Defence
MOG	Mobile Operations Group
51 Mor	51mm light mortar: HE range of 750 metres
81mm Mor	81mm mortar: HE range of 5,650 metres. Also fires smoke and illumination
120mm Mor	120mm heavy mortar used by the Taliban. Range of 7,240 metres
Mousehole	X-shaped charge used to blast a hole through an obstruction
MR2	Nimrod: RAF reconnaissance and surveillance aircraft
NATO	North Atlantic Treaty Organisation
NDS	National Directorate of Security (Afghanistan)
NGO	non-governmental Organisation
Nimrod	RAF MR2 Reconnaissance and surveillance aircraft
OC	Officer Commanding units smaller than a commando/battalion
ODA	Operational Detachment Alpha (US)
OEF	*Operation Enduring Freedom*
OGD	Other Government Departments
OMLT	operational mentoring and liaison team
OP	observation post
Ops One Coy	reserve rifle company at instant notice as a QRF

3 Para	3rd Battalion, The Parachute Regiment
Pathfinder	16 AAB equivalent of the BRF
PDC	Provincial Development Council
PID	positive identification
PJHQ	Permanent Joint Headquarters
PKM	7.62mm GPMG. Range of 1,500 metres
PRR	personal role radio carried/worn by every marine. Range of 500 metres
PRT	provincial reconstruction team
PSCC	Provincial Security Co-ordination Centre
PSO	Peace Support Operations
PSYOPS	Psychological Operations
PVCP	permanent vehicle check point
QIP	Quick Impact Project
QRF	Quick Reaction Force
R&D	research and development
R&R	rest and recuperation Leave (fourteen days after six months in theatre, but including travelling time)
RC(S)	Regional Command (South)
RCIED	radio-controlled improvised explosive device
RIP	Relief in Place
ROC	Rehearsal of Concept (ROC Drill)
107 Rocket	107mm Chinese anti-armour rockets. Range in excess of 7,000 metres
ROE	Rules of Engagement
RPG	rocket-propelled grenade: Russian-made anti-tank missile more often used by terrorists. 90mm calibre with a maximum (aimed) range of 500 metres and capable of penetrating 260mm of armour

RSM	Regimental Sergeant Major; senior non-commissioned officer within a Commando
RSOI	reception, staging and onwards integration
RV	rendezvous
SA-80	5.56mm standard infantry rifle. Optimum range 300 metres, effective to 600 metres
SAF	small arms fire
Sangar	small, defensive position surrounded by a stone or sandbagged parapet
Sappers	nickname for the Royal Engineers
SAS	Special Air Service: UK army special forces
SATCOM	satellite communications
SBS	Special Boat Service: predominantly Royal Marines; special forces
Sec	Section: three or four to a Royal Marine Troop; between six and eight men strong, usually commanded by a corporal
SF	Security Forces
SF	Special Forces
SH	support helicopter
Shura	meeting – usually with the elders of a town or district
SIED	suicide improvised explosive device
SIGINT	Signals Intelligence
SIR	Shooting Incident Report
SITREP	Situation Report
SNCO	senior non-commissioned officer: sergeant, colour-sergeant, warrant officer second class and warrant officer first class
SOP	Standard Operating Procedure
Sqn	squadron

Stripey	Royal Marines sergeant: usually a troop sergeant
TAC HQ	tactical headquarters of a unit, eg Commando Tac HQ
TACP	Tactical Air Control Party
TERP	slang for interpreter
TF	Task Force
TI	thermal imaging
TIC	Troops in Contact
Toms	soldiers in the Parachute Regiment
Tp	troop (Royal Marines equivalent to an army platoon): about thirty strong commanded by a second lieutenant, lieutenant or junior captain
UAV	unmanned aerial vehicle
UKTF	UK Task Force
UOR	urgent operational requirement
VBIED	vehicle-borne improvised explosive device
VCP	vehicle check point
WIA	Wounded in Action
Viking	ATV(P)
WMIK	weapons mount installation kit (fitted to a stripped-down Land Rover)